DISCARD

Nanotechnology Demystified

Demystified Series

Accounting Demystified
Advanced Statistics Demystified
Algebra Demystified
Anatomy Demystified
asp.net 2.0 Demystified
Astronomy Demystified
Biology Demystified
Biotechnology Demystified
Business Calculus Demystified
Business Math Demystified
Business Statistics Demystified
C++ Demystified
Calculus Demystified
Chemistry Demystified
College Algebra Demystified
Corporate Finance Demystified
Databases Demystified
Data Structures Demystified
Differential Equations Demystified
Digital Electronics Demystified
Earth Science Demystified
Electricity Demystified
Electronics Demystified
Environmental Science Demystified
Everyday Math Demystified
Forensics Demystified
Genetics Demystified
Geometry Demystified
Home Networking Demystified
Investing Demystified
Java Demystified
JavaScript Demystified
Linear Algebra Demystified
Macroeconomics Demystified

Management Accounting Demystified
Math Proofs Demystified
Math Word Problems Demystified
Medical Terminology Demystified
Meteorology Demystified
Microbiology Demystified
Microeconomics Demystified
Nanotechnology Demystified
OOP Demystified
Options Demystified
Organic Chemistry Demystified
Personal Computing Demystified
Pharmacology Demystified
Physics Demystified
Physiology Demystified
Pre-Algebra Demystified
Precalculus Demystified
Probability Demystified
Project Management Demystified
Psychology Demystified
Quality Management Demystified
Quantum Mechanics Demystified
Relativity Demystified
Robotics Demystified
Signals and Systems Demystified
Six Sigma Demystified
sql Demystified
Statistics Demystified
Technical Math Demystified
Trigonometry Demystified
uml Demystified
Visual Basic 2005 Demystified
Visual C# 2005 Demystified
xml Demystified

Nanotechnology
Demystified

LINDA WILLIAMS
DR. WADE ADAMS

McGraw-Hill

New York   Chicago   San Francisco   Lisbon   London
Madrid   Mexico City   Milan   New Delhi   San Juan
Seoul   Singapore   Sydney   Toronto

McGraw-Hill books are available at special quantity discounts to use as premiums and sales promotions, or for use in corporate training programs. For more information, please write to the Director of Special Sales, Professional Publishing, McGraw-Hill, Two Penn Plaza, New York, NY 10121-2298. Or contact your local bookstore.

Nanotechnology Demystified

1234567890 CUS CUS 019876

ISBN -13: 978-0-07-146023-1
ISBN -10: 0-07-146023-3

Sponsoring Editor
Judy Bass

Editorial Supervisor
Janet Walden

Project Manager
Seema Koul

Copy Editor
Lisa Theobald

Proofreader
Sunita Dogra

Indexer
Claire Splan

620.5

Production Supervisor
Jean Bodeaux

Composition
TechBooks

Cover Series Design
Margaret Webster-Shapiro

Cover Illustration
Lance Lekander

In Memoriam

This book is dedicated to Richard E. Smalley, Gene and Norman Hackerman Professor of Chemistry and Professor of Physics at Rice University, who had the vision, courage, and quiet persistence to question traditional wisdom, attempt to explain the contradictions in nature and the physical sciences, and seek new solutions to pressing global problems. Great advances in medicine, communications, transportation and energy are sure to come from their efforts.

Dr. Smalley passed away during the final stages of the preparation of this manuscript after a long fight with cancer. His insightful vision will be sorely missed.

L. Williams

ABOUT THE AUTHORS

Linda Williams, M.S., is a nonfiction writer with expertise and experience in the fields of science, medicine, and space. She was a former lead scientist and/or technical writer for NASA, McDonnell Douglas, Wyle Labs, and Rice University. Williams is also the author of *Chemistry Demystified, Earth Science Demystified,* and *Environmental Science Demystified,* all by McGraw-Hill.

Dr. Wade Adams is the Director of the Smalley Institute for Nanoscale Science and Technology at Rice University. He has written more than 190 publications, including several review articles and two edited books.

CONTENTS

PREFACE

Nanotechnology Demystified is for anyone interested in the nanoscale world who wants to learn more about this exciting new area. It can also be used by home-schooled students, tutored students, and those people wanting to change careers. The material is presented in an easy-to-follow way and can best be understood when read from beginning to end. However, if you want more information on specific topics—for example, quantum dots, nanoelectronics, lab-on-a-chip, and so on—or you want to check out only nanotechnology business happenings, those chapters can be reviewed individually.

During the course of this book, I have mentioned milestone theories and accomplishments of many scientists and engineers. I have highlighted these knowledge leaps to suggest how the questions and bright ideas of curious people have advanced humankind.

Science is all about curiosity and the desire to figure out how something happens. Nobel Prize winners were once students who daydreamed about new ways of doing things. They knew that answers to difficult questions had to exist and were stubborn enough to dig for them. The Nobel Prize in science (actors have Oscar and scientists have Nobel) has been awarded more than 470 times since 1901. The youngest person to receive the award, physicist W. Lawrence Bragg, was only 25 years old when he won his Nobel in 1915.

Alfred E. Nobel (1833–1896) held 355 patents for inventions during his lifetime. After his death, his will outlined the establishment of an international annual award in five areas (chemistry, physics, physiology/medicine, literature, and peace) of equal value, "for those who, in the previous year, have contributed best towards the benefits for humankind." In 1968, the Nobel Prize for economics was established. More than 776 Nobel Prizes have been awarded in all areas since the first prize was given out.

Nobel wanted to recognize innovative heroes and reward creative thinking in the quest for knowledge. My hope is that by describing some of the discoveries changing our understanding of how things work, you'll focus your own creative energy toward tackling important science and engineering questions.

This book provides a general nanotechnology overview with sections on all the main areas you'll find in a nanotechnology class or an individual study of the subject. The basics are covered to familiarize you with the terms, concepts, and tools most used by nanoscience/ nanotechnology researchers and engineers. I have listed helpful Internet sites that include up-to-date and fascinating new methods and information.

Throughout the text, I have supplied illustrations to help you visualize what is happening on the nanotechnology front. You'll also find quiz, test, and exam questions throughout the book. All the questions are multiple choice and much like those used in standardized tests. A short quiz appears at the end of each chapter. These quizzes are "open book," so they should be fairly easy. You can look back at the chapter text to refresh your memory or check the details of a natural process. Write down your answers and have a friend, parent, or tutor check your score with the answers in the back of the book.

This book is divided into four major parts. A multiple-choice test follows each of these parts. When you have completed a section, you can take the accompanying test. Take the tests "closed book" when you are confident about your skills on the individual quizzes. Try not to look back at the text material during the test. The text questions are no more difficult than those of the quizzes, but they serve as a more complete review. I have thrown in lots of wacky answers to keep you awake and make the tests fun. A good score is 75 percent or better correct answers. Remember that all answers are located in the back of the book.

The final exam at the end of the course comprises questions that are easier than those of the quizzes and tests. Take the exam when you have finished all the chapter quizzes and part tests and feel comfortable with the material as a whole. A good score on the final exam is at least 75 percent correct answers.

With all the quizzes, tests, and the final exam, you may want to have a friend, parent, or tutor tell you your score without telling you which of the questions you missed. Then you will not be tempted to memorize the answers to the missed questions, but can instead go back and see if you missed the point of the idea. When your scores are where you'd like them to be, go back and check the individual questions to confirm your strengths and any areas that need more study.

Try reading through a chapter a week. An hour a day or so will allow you to take in the information slowly. Don't rush; just plow through at a steady rate. Nanotechnology is not difficult, but the topic does involve some thought in deciphering some of its implications.

You may want to linger in a chapter until you have a good handle on the material and get most of the answers correct before moving on to the next chapter. If you are particularly interested in public policy, spend more time reviewing Chapter 11. If you want to learn the latest about how nanomaterials may be used in environmental remediation, allow more time to study Chapter 10.

After completing the course and becoming a "nanotechnologist-in-training," this book can serve as a ready reference guide with its comprehensive index, appendices, and examples of nanocrystalline types, biological markers, and potential for quantum computing.

Linda Williams

ACKNOWLEDGMENTS

Illustrations in this book were generated with Microsoft PowerPoint and Word courtesy of Microsoft Corporation.

National Nanotechnology Initiative (NNI), Office of Research and Development (ORD), Environmental Protection Agency (EPA), and other governmental agency information has been used as indicated.

A very special thanks to Kristen Kulinowski, Ph.D. (Rice University, faculty fellow, Executive Director of Education & Policy at the Center for Biological and Environmental Nanotechnology) for the technical review of this book and to the Rice University faculty who provided research images for this work.

Thank you to Wade Adams, Ph.D., director of the Smalley Institute for Nanoscale Science and Technology for nanotechnology history and topical discussions.

Many thanks to Judy Bass at McGraw-Hill for her amazing energy and support despite unseen hurdles and life's intrusions.

Elisabeth, Paul, Bryn, Evan, and Jack—thank you for your love and encouragement.

Linda Williams

PART ONE

Discovery

CHAPTER 1

Buckyball Discovery

Every once in awhile (from centuries to millennia), something new is discovered or created that changes *everything*. Cave dwellers smelled something new (smoke) and decided (after a long committee meeting) to check it out. Fire changed everything. Sushi was out and barbeque was in.

Skip forward a bit to a time when some inventive artisans figured out how to make tools out of iron. These sturdy tools lasted a lot longer than their stone implements and gave some people the idea that taking over the world might be an interesting endeavor.

Fast forward again, to a time of electricity, horseless carriages, antibiotics, and indoor plumbing for the majority of the developed nations. Suddenly, the human race became aware that if something was considered long enough (out of committee), anything was possible. Science and technology breakthroughs exploded in ways that were laughed at as pure fiction only decades earlier.

Then, the true age of technology—color TV and fast computers—emerged. "Smaller, faster, lighter, and smarter" became the anthem of the day. The more we knew, the more we wanted to learn. We hungered for knowledge of how things worked. Our curiosity was limitless. Everything from quasars and plate tectonics to DNA and dung beetles captured our interest.

Today, the quest for knowledge has reached a fevered pitch. A new, what-do-you-know, gee-whiz, fantastical, mind-blowing, paradigm-changing ability has been discovered: *nanotechnology*.

The reality of nanotechnology is evolving in ways that give us everything from faster and tinier computers, better tennis balls, and stain-resistant clothing, to transparent sunscreens (SPF60), molecular sensors, and cell-specific cancer therapies. Today, hundreds of products on the market use nanotechnology. Most of these products are the result of better uses of established technology, such as scratch-resistant, anti-adhesive coatings, but in the next 10 to 20 years, emerging technologies will knock our socks off.

Throughout the course of this book, nanomaterials, nano-applications, and many of the amazing technologies on the nanoscale horizon will be described. Additionally, nanotechnology's impact will be examined from various angles of investment opportunities, products, risk, public policy, and international impact.

So sit back, put your feet up, and get ready to enter a world of the super small, the world of imagination come true, the world of nanoscience and nanotechnology.

In the Beginning

In 1897, J.J. Thomson discovered negatively charged particles by removing all the air from a glass tube that was connected to two electrodes. His *cathode ray tube* (CRT) used a current to excite atoms of different gases contained in the tube. Electricity was beamed directionally through the tube from one electrode to the other (electrode). By using this tube, scientists of a century ago began to separate the individual particles that make up atoms.

ELECTRONS

Through his early experiments with several different colored gases, Thomson found that *electrons* (−) had a negative charge and seemed to be common to all elements. This was exciting news, since most people considered the differences between elements to be pretty mysterious.

> *Electrons* are small, negatively charged subatomic particles that orbit around an atom's positively charged nucleus.

In 1906, Thomson was awarded the Nobel Prize in physics for this research and his electrical work with gases. Later research found that an electron has a mass of 9.1×10^{-31} kg and that it has a charge of 1.6×10^{-19} Coulombs.

NUCLEUS

It wasn't until scientists discovered that the atom was not just a solid chunk, but was in fact made up of smaller subparticles located in and around a nucleus, that even more questions were asked.

In 1907, a student of Thomson named Ernest Rutherford developed the modern atomic concept. He received the Nobel Prize for chemistry in 1908 and was knighted in 1914 for his work. (Who said chemistry was not a glory science?) Through his experiments with radioactive uranium in 1911, Rutherford described a *nuclear model*. By bombarding particles through thin gold foil, he predicted that atoms had positive cores that were much smaller than the rest of the atom. His experiments, along with those of his student, Hans Geiger (of Geiger counter fame), showed that more than 99 percent of the bombarded particles passed easily through the gold, but a few (1/8000) ricocheted off at wild angles, even backward. Rutherford thought this scattering took place when (+) nuclei of the test particles collided and were then repelled by heavy positively charged gold nuclei. It was later proven that when an accelerated particle collided with an electron of a gold atom in a gas, a proton was knocked out of the nucleus.

Later research, conducted along the same lines as Rutherford's early work, found that each proton in a nucleus has a mass of more than 1800 times that of an electron. In fact, the positively charged atomic nucleus contained most of its mass. The nucleus was very dense and took up only a tiny part of an atom's total space. Figure 1-1 shows the basic atomic structure that Rutherford predicted. Later research showed that electrons don't actually orbit the nucleus like planets around the sun, but are more like a cloud of mist swirling around the nucleus.

To get an idea of scope, picture an atomic nucleus the size of a ping-pong ball. The rest of the atom, with its zippy, circling, negatively charged electrons, would

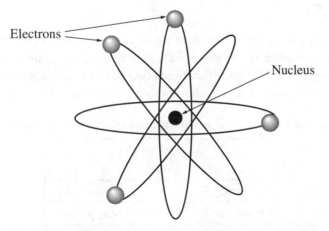

Figure 1-1   Rutherford's early concept of atomic structure.

measure nearly 3 miles across. More accurately, nuclei are roughly 10^{-12} meters in diameter in the nano world!

Protons

A *proton* is a smaller bit of matter or *subatomic particle* within the nucleus. As mentioned, a proton has a positive charge and roughly 1800 times greater mass than an electron. The *atomic number* (Z) of an element is derived from the number of protons in an atom's nucleus. A *pure element* is one that is made up of particles that all have the same atomic number.

Neutrons

The nucleus of an atom contains subatomic particles called *nucleons*. Nucleons are divided into two kinds of particles called *neutrons* and *protons*. Protons make up the dense nucleus core, but when chemists made calculations based on atomic weights of atoms, the numbers didn't add up. They knew something was missing; then neutrons were discovered.

> *Neutrons* are subatomic particles with a similar mass to protons, but no electrical ($+$ or $-$) charge. They are neutral.

Neutrons are nuclear particles that have no charge and are located inside the crowded nucleus with positively charged protons. To give you a better idea of how they compare in relation to size, Table 1-1 lists common characteristics of super small electrons, protons, and neutrons.

MOLECULES

Though many forms of matter, such as wood, rock, or soap, appear solid upon first inspection, most matter is composed of a combination of atoms in a specific geometrical arrangement. The force that binds two or more atoms together is known as a *chemical bond*. A *molecule* is the basic joining of two or more atoms held together

Table 1-1 Electrons, protons, and neutrons are nanoparticles with various characteristics.

Atomic Particles		
Name	*Symbol*	*Mass (g)*
Electron	e$^-$	9.110×10^{-28}
Proton	p$^+$	1.675×10^{-24}
Neutron	n	1.675×10^{-24}

by chemical bonds. In a *covalent bond,* electrons are shared, while in an *ionic bond,* electrons are transferred.

> A *molecule,* the simplest structural unit of an element or compound, is composed of atoms chemically bonded by attractive forces.

One familiar compound is composed of two atoms of hydrogen and one atom of oxygen. This compound, water, is held together by covalent bonds. Compounds written with two or more different element symbols are called *formulas* of the compound. The formula for water is H_2O. The number of atoms of each element is written as a subscript in the formula—the *2* in H_2O. When no subscript is included, it is understood that only one atom of the element is involved.

In a molecular substance, the molecules are all alike. The molecules are so small that even extremely small samples contain huge numbers of molecules. For example, a raindrop of about 5 mm in diameter (a bit smaller than a quarter of an inch) contains about 2×10^{21} molecules, which is about 2000 billion billion molecules (sometimes called 2 *sextillion* in English)! To show you how big that number is, if each water molecule were as thick as the piece of paper in this book, and you stacked up 2×10^{21} pieces of paper, that stack would reach from the earth to the sun (91 million miles) and back—about 600,000 times! Or, if each water molecule (around 0.3 nanometers in diameter) could be lined up in a string, it would reach from the Earth to the sun and back, twice!

Listing subscript numbers is important when researchers want to combine or separate different compounds. The following examples show a few other molecular ratios in different compounds. Some simple compound formulas are listed here:

- Sodium chloride (NaCl) = 1 sodium atom and 1 chlorine atom
- Hydrogen peroxide (H_2O_2) = 2 hydrogen atoms and 2 oxygen atoms
- Ethanol (C_2H_6O) = 2 carbon atoms, 6 hydrogen atoms, and 1 oxygen atom

Since the Earth contains many different forms of matter, solid, liquid, and gas, it is easy to see that atoms can combine at the nanoscale in nearly infinite ways to form molecular compounds. However, only a certain number of discovered elements exists, and sometimes chemical formulas are the same for different compounds. The way chemists keep these formulas straight is through their *molecular* and *structural* formulas.

Molecular and Structural Formulas

A *molecular formula* is more specific than a compound's name. It provides the exact number of different atoms of each element in a molecule. We saw this earlier in the formula for water, H_2O. Think of the molecular formula as a closer look, like being shown the difference between a long-bed truck and an 18-wheel truck/trailer combination. The components are basically the same—engine, tires, body, and frame—but the number of wheels and length of the vehicle can make all the difference in its size and function.

A specific molecule is always composed of the same number and kinds of atoms, chemically bonded by attractive forces. These atoms are usually held together in a certain way. This bonding comes about because of electron properties and their location around each atomic nucleus.

Table 1-2 Molecular formulas provide the number of elemental atoms in a compound

Name	Chemical Formula
Ammonium carbonate	$(NH_4)_2CO_3$
Ammonium nitrate	NH_4NO_3
Benzene	C_6H_6
Calcium hydroxide	$Ca(OH)_2$
Carbon tetrafluoride	CF_4
Cinnemaldehyde	C_9H_8O
Cupric nitrate	$Cu(NO_3)_2$
Dichlorodiphenyltrichloroethane(DDT)	$C_{14}H_9Cl_5$
Diphosphorus trioxide	P_2O_3
Fluoromethane	CH_3F
Fructose	$C_6H_{12}O_6$
Ethane	C_2H_6
Gallium oxide	Ga_2O_3
Lithium dichromate	$Li_2Cr_2O_7$
Magnesium chloride	$MgCl_2$
Oxalic acid	$H_2C_2O_4$
Peroxide	H_2O_2
Potassium nitrate (saltpeter)	KNO_3
Sodium chloride	$NaCl$
Sodium stearate	$C_{18}H_{36}O_2Na$
Sulfuric acid	H_2SO_4
Urea	$CO(NH_2)^2$

A simple molecular formula such as $CuSO_4$ (copper sulfate) tells us the number of copper, sulfur, and oxygen atoms of the different elements in the sample. In Table 1-2 you can see some common molecular formulas.

> A *molecular formula* provides the exact number of atoms of each element in a molecule.

Water is written as H_2O, saltpeter (Potassium nitrate, used in fireworks and fertilizer) is KNO_3, and fructose (the sugar found in fruit and honey) is $C_6H_{12}O_6$.

> A *structural formula* shows how specific atoms are ordered and arranged in a compound.

Structural formulas show every atom and every bond. Atoms are represented by their atomic symbol, and bonds are shown by solid black lines. A single line represents two shared electrons in a single covalent bond. Two lines represent four shared electrons in a double covalent bond.

> Any two molecules with the same molecular formula but a different arrangement of molecular groups are called *isomers*.

Figure 1-2 shows the molecular formula C_2H_6O and two isomers with different structures and functional groups.

A structural formula shows exactly how an element is connected to the others in the molecule. You can think of a structural formula as being like a football game: The plays are set up with different players placed in certain positions. Each play is designed to serve a particular purpose. If the players form up one way, the quarterback may throw the ball. Set them up another way and the end player runs the ball.

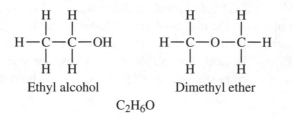

Ethyl alcohol   Dimethyl ether

C_2H_6O

Figure 1-2   Isomers have the same molecular formula and different structures.

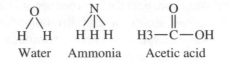

Figure 1-3   Structural formulas show the arrangement of atoms in a molecule.

If the players on the other side don't react to a certain configuration in the predicted way, the quarterback may have to run the ball. Placement and function of individual players is everything in football, and the same is true of chemistry. The structural arrangement of the atoms in a molecule can make a big difference in the characteristics and reactivity of compounds. Figure 1-3 shows structural formulas with individual elements indicated.

Researchers study the structure of a molecule to figure out how it will react in a reaction. Structure has a definite effect on the properties of nanoparticles.

Plenty of Room at the Bottom

On December 29, 1959, Professor Richard Feynman (a 1965 Nobel Prize winner in physics) presented a lecture entitled "There's Plenty of Room at the Bottom" during the annual meeting of the American Physical Society at the California Institute of Technology (Caltech). He described a field that few researchers had thought much about, let alone investigated. Feynman presented the idea of manipulating and controlling things on an extremely small scale by building and shaping matter one atom at a time.

He amazed his audience with an idea so simple it was outrageous (at least at the time and with the tools available). Feynman pointed out that some scientists thought most of the big discoveries had been made and that science just wasn't that exciting anymore. He went on to show how they were wrong.

He described how the 24 volumes of the *Encyclopedia Britannica* could be written on the head of a pin. He imagined raised letters of black metal that could be reduced to 1/25,000 of their normal size (the size of this type). Feynman discussed how such a work could be read using an electron microscope in use at that time. The trick, he said, was to write the super small texts and scale them down without loss of resolution.

How is it done? Feynman said that letters could be represented by six to seven bits of information for each letter. He also suggested using the inside as well as the surface of a metal to store information. Feynman allowed that if each bit was equal to 100 atoms, all the information of all the books in the world could be written in a cube of material 1/200 of an inch wide, about the size of a tiny speck of dust. There *is* plenty of room at the bottom!

Feynman (a physicist) insisted that this was old news to biologists. Biologists had studied cell proteins such as DNA (deoxyribonucleic acid) molecules for decades. Scientists knew that DNA located in the nucleus (mission control) of cells encodes for the design of everything from a gnat to a human to a killer whale. And everything in between!

Feynman suggested that biologists had just been waiting for physicists to design a microscope that was 100 times more powerful. Once they had more powerful tools, scientists would have a window into protein interactions up close and personal. Feynman talked about the countless possibilities of the molecular world; now called the "nano world." He ignited his colleagues' imaginations as well as a scientific race to explore and characterize this molecular world.

But Feynman's topic was not entirely new. The idea of changing a chemical's properties first began with the early alchemists. Those early scientists looking for an immortality elixir or a "get rich quick" formula of changing lead into gold knew that chemical purification and reactions could make things happen. They were actually trying to do nanotechnology by combining atoms in certain ways to get desired compounds.

In 1981, Gerd Binnig and Heinrich Rohrer of IBM's Zurich Research Laboratory created the scanning tunneling microscope that allows scientists to see and move individual atoms for the first time. They found that by using an electrical field and a special nanoprobe with a super small tip, they could move atoms around into forms that they wanted. Since then, the scanning tunneling microscope has led to the development of the atomic force microscope, one of the advanced measurement tools of the nano era. New ideas about matter were born. This was a really significant invention since it was the first time that individual atoms could be imaged and manipulated. Rohrer and Binnig won a the Nobel Prize in physics in 1986 for their design of the scanning tunneling microscope.

In 1989, Don Eigler at the IBM Almaden Research Center in San Jose, California, formed the letters *IBM* from 35 xenon atoms and photographed his success.

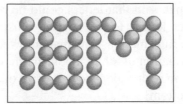

Figure 1-4   Scientists showed how atoms could be moved individually.

Figure 1-4 demonstrates how individual atoms were used to form the letters *IBM*. You will read more about the tools of nanotechnology in Chapter 4.

LUCK AND INSIGHT

In September 1985, a new kind of carbon (C_{60}) was discovered by three innovative chemists—Robert F. Curl Jr., Sir Harold W. Kroto, and Richard E. Smalley, who came together at Rice University in Houston, Texas, to perform a set of experiments that changed chemistry and the world. They were assisted by two graduate students, James Heath (now a chemistry professor at the California Institute of Technology) and Sean O'Brien (now a scientist with Texas Instruments in Dallas), who helped perform some of the experiments. Since a single Nobel Prize can be divided among up to three people only, the graduate students shared the historical spotlight but not the prize awarded on December 10, 1996 (the 100th anniversary of Alfred Nobel's death).

The new carbon family was named the *fullerenes*. The fullerenes—soccer ball shaped, cage-like molecules, characterized by the symmetrical C_{60}—soon occupied center stage in chemistry. Very different from known carbon forms like graphite and diamond, C_{60} (made up of 60 carbons) was officially named *Buckminster fullerene* (in honor of architect and inventor R. Buckminster Fuller, who designed and built the first geodesic dome). Figure 1-5 shows the basic C_{60} (buckyball) soccer ball shape.

The actual discovery of fullerenes came about through Smalley and Kroto's experiments on an instrument Smalley invented to study molecules and clusters of atoms. Kroto was interested in Smalley's laser vaporization technique to verify a theory he had about the carbon thrown off by long-chain carbons in interstellar space. Kroto thought that carbon-rich red giant stars were giving off complex carbon species that radio astronomy should be able to detect.

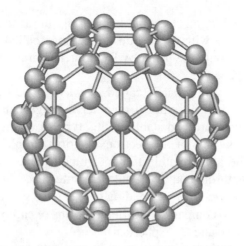

Figure 1-5   A Buckmister fullerine is shaped like a multifaceted soccer ball.

The research group tried to figure out the structure for the carbon's unique chemical signature using an instrument called a *mass spectrometer* (which measures the wavelengths and energies of elements). It finally came together late one night when Smalley pieced together a construction paper and adhesive tape polygon that had the all important 60 vertices in a highly symmetrical closed shell. This new carbon molecule (C_{60}) was nicknamed the *buckyball*. While graphite contains carbon atoms formed in flat sheets, buckyballs are open spherical cages with strong carbon-to-carbon bonds.

Everyone who thought carbon existed only as graphite and diamond couldn't believe it. Was it a mistake? Why hadn't they seen this new carbon group before? How important could it be? Many scientists started studying this new family of molecules, and the importance of the buckyball molecule was clear to everyone when Smalley, Curl, and Kroto won the 1996 Nobel Prize in chemistry for their amazing discovery. Because of his ability to speak out for research on buckyballs and fullerenes, and later for all of nanotechnology, Smalley has often been called one of the fathers of nanotechnology, along with Binnig and Rohrer. Feynman is often called the grandfather of nanotechnology.

GRAPHITE

Before fullerenes were identified, graphite was probably the most understood of the complex carbons. Graphite has a layered or planar (flat) structure. The carbon structure is complex, but mostly two dimensional (2-D) in a flat plane—think chicken wire. Or you can think of graphite as being like a flat playing card: On edge, a card is fairly strong and flat and will slide over other cards, but it will bend or break if too much force is used.

Soft, light, and flexible, most people know graphite as the black stuff used in pencils. When you write with a graphite (sometimes called lead) pencil, bits of graphite are rubbed off onto the paper by friction. The graphite sticks to the paper to form letters or shapes.

Graphite molecules have what is called *covalent bonding*. Covalent bonding holds hard solids together. Assembled together in large nets or chains, covalent multi-layered solids are hard and stable in this type of configuration. In graphite, each carbon atom uses three of its electrons to form simple bonds with its three closest neighbors. The atoms within the flat graphite layers are held together by strong covalent bonds throughout the whole graphite sheet. Other attractive forces, called *van der Waals forces,* provide bonding between the sheets to hold the graphite solid together. Figure 1-6 compares the different forms of carbon—graphite (planar), diamond (crystal lattice), and C_{60} (spherical).

Graphite feels like a slippery powder and is used as a dry lubricant for locks and athletic equipment. Its bonding arrangement gives it useful properties such as a high melting point. Graphite's entire bonding structure has to be broken for it to melt. Graphite is insoluble in water and organic solvents, but it conducts electricity. So forget about sticking a pencil in a light socket! You could make a shocking discovery!

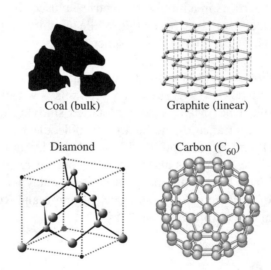

Coal (bulk)   Graphite (linear)

Diamond   Carbon (C_{60})

Figure 1-6   Graphite has flat layers, while diamond and C_{60} have more complex shapes.

DIAMOND

Sparkling diamonds, the strongest molecules known before C_{60} was discovered, are the favorites of brides as well as industrial engineers. Diamond is the third form of carbon. Through covalent bonding, diamond atoms are arranged into 3-D solids, in which one carbon atom is covalently bonded to four other carbons. A diamond's 3-D crystal structure makes it super hard for cutting or grinding through automotive steel and other tough manufactured materials. In fact, diamond is the hardest known solid.

Diamond also has a very high melting point (almost 4000°C) due to its super strong carbon-carbon covalent bonds that have to break throughout the structure before it will melt. Diamond is also insoluble in water and organic solvents. The bonds are just too strong between the covalently bonded carbon atoms.

Unlike graphite, diamond doesn't conduct electricity. All the electrons are held tightly between the atoms and can't move easily throughout the solid. (Think of Times Square in New York City on New Year's Eve. Standing room only.)

Cut diamonds, of great brilliance and luster, have been considered a treasure for centuries. But to scientists, diamonds are important for their range of extraordinary and extreme properties. When compared to almost any other material, diamond is pretty much the champion. As well as being the hardest known material, it is the stiffest and least compressible. Diamonds are also the best thermal conductors with extremely low thermal expansion. They don't react to most strong acids or bases. Diamonds are clear from deep ultraviolet light through the visible wavelength range to the far infrared spectrum.

Graphite vs. Diamond vs. C_{60}

A lot of the minerals are made up of only a single element. Geologists sometimes subdivide minerals into *metal* and *nonmetal* categories. Of all elements, 80 percent are metals. Gold, silver, and copper, for example, are metals. As we know, carbon makes up the minerals graphite, diamond, and C_{60}, which are nonmetals.

Diamond's arrangement of carbon atoms in a lattice gives it amazing properties. So how do graphite, diamond, and C_{60} compare? All are made up of carbon. In diamond, we have the hardest known material; in graphite, one of the softest. The big difference is the way the atoms are bonded together. In diamond, each carbon is bonded to four others, while graphite has only three other bonds, in sheets of connected benzene (six carbon) rings, bonded to each carbon. Because the sheets can slide over one another, graphite is slippery.

Fullerenes are like diamond and graphite, only better. They have some of the characteristics of both, but "kicked up a notch." Since about 1990, research on how C_{60} compares to other carbon forms really took off. Fullerenes with 70 (slightly

oval), 80 (sausage shaped), and even more carbons were discovered. Each new discovery raised new questions about properties and functions. As more forms and combinations were found, scientists also discovered that electron energy and electrical currents acted differently at the molecular level. The rules by which graphite and diamond had to play didn't seem to affect the fullerences and better, improved instrumentation paved the way for even more discoveries.

Single-Walled Carbon Nanotubes

In 1991, *carbon nanotubes* were discovered in soot on a carbon rod arc cathode by Sumio Iijima at NEC Fundamental Research Laboratories in Tsukuba, Japan. Iijima's high-resolution multi-walled carbon nanotube (MWNT) electron micrographs illustrated that the new carbon species with rounded end caps were fullerene cousins. But while MWNTs are related to fullerenes, they were not molecularly perfect.

However, the single-walled carbon nanotubes (SWNTs) discovered in 1993, simultaneously by Iijima and Toshinari Ichihashi at NEC in Japan and Donald S. Bethune and others at IBM Almaden Research Center in San Jose, California, were different. The two groups described how iron, nickel, or cobalt, inserted into the anode of a carbon arc and run to produce C60, yielded a rubbery type of soot on the chamber walls. Transmission electron microscopy (TEM) soot images showed that the soot was made up of many SWNTs with a narrow distribution of diameters. The soot contained no MWNTs.

The puzzles and process of fullerenes discovery continue today. Thousands of scientists and engineers are working on fullerene chemistry and physics to make larger amounts of material, to make them in pure forms, and to study their properties. Figure 1-7 illustrates a carbon nanotube.

Nanotubes contain thousands to millions of carbon atoms, depending on their length. Nanotubes can have metallic properties comparable to or better than copper, or they can be semiconductors, such as silicon in transistors, depending on their structure. They can conduct heat as well as diamond, and because they are carbon, a chemist can create bonds between the fullerene carbon atoms and other atoms or molecules. This ability to attach other molecules to buckyballs or nanotubes makes them a new nanomaterial to use with biological systems or to bond into composite materials. The theoretical scientists calculate that nanotubes will be able to make the strongest fibers ever made (about 100 times stronger than steel), with only 1/6 the weight. Carbon buckyballs and nanotubes are the most exciting new material discovery in many decades!

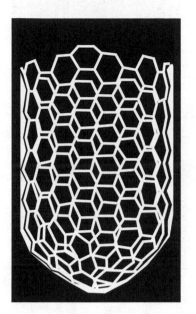

Figure 1-7   Carbon nanotubes have a symmetrical structure.

FUTURE NANOTECHNOLOGISTS

The training of undergraduate, graduate, and postdoctoral candidates is important to nanotechnology in many areas, including biology, chemistry, physics, materials science, chemical engineering, applied physics, computer science, and electrical engineering. Many nanotechnology disciplines are developing and will need trained people, but ceramics, polymers, semiconductors, metal alloys, catalysts, and sensors are thought to have a growing employment window. You'll read more about these in Chapter 8.

Nanotechnology's interdisciplinary nature requires that a student have both a solid grounding in a single discipline as well as the ability to think across disciplines. It's possible that future nanotech discoveries (such as medicine and nanocomposites) will come about through entirely new types of curricula not bound by traditional disciplinary limits.

Let's Roll

The world of the nanoscale touches every area. Figure 1-8 chronicles the major nano events that got us where we are today.

Smaller, faster, lighter, and smarter brought us to the cusp of nanotechnology breakthroughs that have the potential to benefit all of humankind. Imagination and

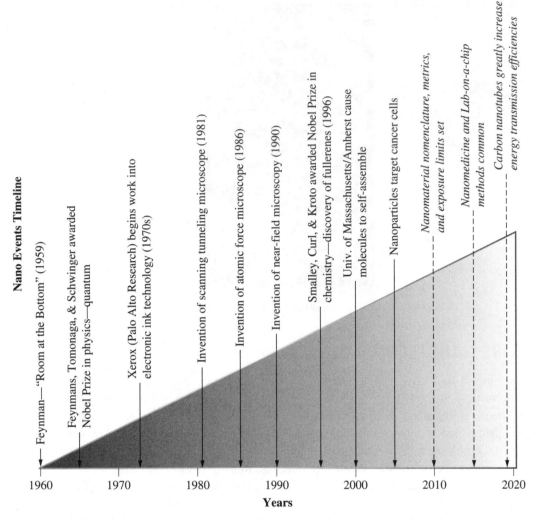

Figure 1-8   The growth of nanotechnology.

function are becoming reality. Today we are fast approaching a chance to touch the future in ways that haven't made this much difference since iron (versus bronze) was the hot topic around the campfire. In the coming chapters, we will explore what all the excitement is about. Get ready. Buckle up and hang on!

Quiz

1. The new carbon family discovered in 1985 was named
   (a) inert gases
   (b) lanthanides
   (c) Rare Earth
   (d) fullerenes

2. The physics lecture entitled "There is Plenty of Room at the Bottom" describing the nanoscale was given by
   (a) Richard Smalley
   (b) Richard Burton
   (c) Richard Feynman
   (d) Richard Petty

3. Nanotubes can have metallic properties, comparable to
   (a) brass
   (b) lead
   (c) tin
   (d) copper

4. Some fullerenes are the shape of which of the following?
   (a) anvil
   (b) sausage
   (c) eggplant
   (d) icicle

5. The mass of a proton in the nucleus is how many times greater than that of an electron
   (a) 800
   (b) 1200
   (c) 1600
   (d) 1800

6. What encodes for the design of everything from a gnat to a killer whale?

   (a) Morse code

   (b) radio waves

   (c) DNA (deoxyribonucleic acid)

   (d) MEMS (microelectromechanical systems)

7. In diamond, each carbon is bonded to how many other atoms?

   (a) 2

   (b) 4

   (c) 6

   (d) 8

8. What carbon types are related to fullerenes, but lack perfect symmetry?

   (a) amino acids

   (b) carbohydrates

   (c) multi-walled nanotubes

   (d) black ash

9. The 1996 Nobel Prize in chemistry was awarded for the discovery of

   (a) quartz

   (b) fullerenes

   (c) uranium

   (d) polonium

10. Researchers use what element to get carbon to form nanotubes?

    (a) iron

    (b) uranium

    (c) potassium

    (d) beryllium

CHAPTER 2

Nanoscale

Is seeing really believing?

Olympic athletes make difficult sports look easy. Magic tricks make the impossible seem possible. Movie special effects make the imaginary believable. So can you really believe your eyes? Scientists and engineers believe what they see only after careful testing and retesting. They realize that whether they're solving planetary mysteries or studying a bat's chromosomes, anything is possible if you understand how it works.

Today, technology is racing forward so fast that the microscopic scale is no longer cutting edge. A microscopic rendezvous with a dust mite, big news 30 years ago, is now ho-hum. Strange, unseen worlds only imagined in the past are coming into sharp focus through extreme science, complex instruments, and dynamic engineering.

Nano is the cool thing now. Like space travel and the Internet before it, the possibilities of the nano world catch the imaginations of school children and scientists alike. Nanotechnology is also an important topic for financial investors. New technology usually means new products and new ways to back a profitable product. Investors and policy makers have a lot of interest in the nanoscale.

Micro vs. Nano

A short time ago, micro was the king of small. From microphones and microwaves to microscopes, microorganisms, and microspheres, everything micro was better. Since their invention, cell phones and computer chips continue to get smaller. Small is good to a point, but today's microelectronics have nearly reached the end of an ever-shrinking micromanufacturing path. Most microscale design has gone almost as far as it can go. To get more speed and power, computer electronics are reaching down into the nanoscale. Computer circuits have gotten so small that they heat up when operating and can burn themselves out. Something has to give.

Nano to the rescue! *Nanoscience* is the study of a seriously small world—the world of atoms and molecules.

A ***nanometer*** is one billionth (10^{-9}) of a meter.

Particles are considered to be *nanoparticles* if one of their dimensions is less than 100 nanometers (nm) across. The prefix *nano* means one billionth. Table 2-1 gives you an idea of how distances and sizes compare. For example, if a gold nanoshell (a nanoparticle that you will meet in Chapter 6) were the size of a marble, an average human would be the size of Mt. Everest (29,035 feet)!

Table 2-1   Comparison of objects and distances

Sample	Measurement (meters)
Uranium nucleus (diameter)	10^{-13}
Water molecule	10^{-10}
DNA molecule (width)	10^{-9}
Protozoa	10^{-5}
Earthworm	10^{-2}
Human	2
Mount Everest (height)	10^{3}
Earth (diameter)	10^{7}
Distance from the Sun to Pluto	10^{13}

An average molecule can comprise between 1 to 25 atoms, giving it a radius of less than 1 to about 10 nm. A molecule, by definition, must have more than one atom. The smallest molecule is H_2. Some *biomolecules* are much, much larger than this, such as DNA (deoxyribonucleic acid). A nanoparticle contains between 50 and 200,000 atoms, so its dimensions can be from a few nanometers to several hundred nanometers. An average bacterial cell, for example, measures a few hundred nanometers across, and a red blood cell is about 6000 nm wide. The smallest parts of a microchip today are around 130 nm across.

New and emerging technology as well as the ability to manipulate atoms and molecules and nanoparticles will allow us to build or change the structures of everyday things—from cancer cells to nanocomputers; everyone wins when such new discoveries are made. The big difference between using nanotools (nanotweezers, optics, magnetics, and electricity) and regular lab stuff (beakers and Bunsen burners) is size, or scale.

Guinness World Record Nanotube

The smallest things that the human eye can make out are around 10,000 nm across—like the tiniest circuits on a motherboard. No one can see carbon nanoparticles and nanotubes without high-powered or scanning probe electron microscopes, so it's tough to understand just how small they really are.

To help everyone understand the nanoscale and carbon nanotubes, some faculty, staff, and more than 100 students at Rice University decided to do something big with nano. They decided to make the invisible visible by building a really big single-walled carbon nanotube (SWNT) model.

On Earth Day, April 22, 2005, using 65,000 bright blue plastic pieces from chemistry model kits, the group at Rice built the world's largest carbon nanotube model. The large-scale model of a 0.7 nm wide, 700 nm long nanotube stretched as long as Houston's tallest skyscraper is high and set the Guinness Record for the world's largest nanotube model at 1180 feet long. If they had built a model of an actual nanotube measuring 5 cm long, the model would have stretched for 15,000 miles! Figure 2-1 shows the SWNT model that the Rice students built.

That day, not only scientists and engineers, but everyone could see how something seriously small could be *really big*! Pieces of the model are on permanent display at the Houston Museum of Natural Science to teach students and adults alike the structure and relative size of the "next big thing."

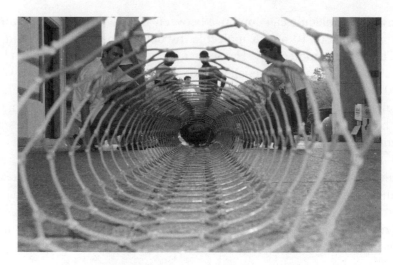

Figure 2-1   Guiness Record for the world's largest nanotube (Courtesy Rice University).

Size Matters

In Greek, *nano* means dwarf, but in science, *nano* means 1 billionth or 1×10^{-9}. So if you are talking about 1 billionth of a meter, you're talking a nanometer. A nanosecond, one billionth of a second, is incredibly fast. A beam of light travels about one foot in a nanosecond.

The metric system, suggested by Gabriel Mouton in 1670, was adopted by the French government as the standard unit of measure in 1795. The metric system is based on the meter and kilogram. The metric system is a decimal system in which all units are increased by multiples of 10. We know that the meter is equal to about 40 inches (a bit longer than a yard), and the kilogram weighs just over 2 pounds.

Table 2-2 shows the prefixes used to describe metric system units.

Nanoscale particles are smaller than can be seen. But is nanotechnology real? How do we know if things that can't be seen are really there at all? Scientists around the world are checking it out.

> A ***nanometer*** is equal to 1 billionth of a meter. An average human hair is about 80,000 nanometers wide.

Scientists discovered that at the nanoscale, size matters! Nanoparticles exist at the size of single atoms, which are about 0.1 nm wide. So, 1 nm is almost equal to 10 hydrogen atoms, stretched end to end. (Remember that hydrogen is the smallest atom.) This subatomic world can't be seen by the naked eye because it's super small!

Table 2-2 Most scientific data is reported in terms of metric measurements

Metric Prefix	Multiples of Ten	Number	Name
exa	10^{18}	1,000,000,000,000,000,000	quintillion
peta	10^{15}	1,000,000,000,000,000	quadrillion
tera	10^{12}	1,000,000,000,000	trillion
giga	10^{9}	1,000,000,000	billion
mega	10^{6}	1,000,000	million
kilo	10^{3}	1000	thousand
	10	1	
milli	10^{-3}	1/1000	thousandth
micro	10^{-6}	1/1,000,000	millionth
nano	10^{-9}	1/1,000,000,000	billionth
pico	10^{-12}	1/1,000,000,000,000	trillionth
femto	10^{-15}	1/1,000,000,000,000,000	quadrillionth
atto	10^{-18}	1/1,000,000,000,000,000,000	quintillionth

Size matters because the properties of nanomaterials can be uniquely different from properties of materials in bigger bulk forms. There are two reasons for differences in a material's nanoscale behavior. First, nanoparticles have a much great surface area per unit volume—that is, a hunk of metal, for example, has much more surface area when broken down into tiny particles than it does when whole. Since chemistry of solids occurs at these surfaces, more surfaces mean increased chemical reactivity. Second, the smaller the particles get, the greater the changes in the particles' magnetic, optical, and electrical properties.

Nanoparticles are party animals, because they like to interact with others without the extra baggage of size to slow them down. Size also affects properties such as color. Nanoparticles of different sizes and shapes can provide a rainbow of different colors. All these differences allow for new application opportunities.

Nanoparticles couldn't be seen earlier because the tools needed to find, study, and control such small structures weren't invented. Today, we see much more than early scientists were able to see. High-resolution microscopes and other instruments allow scientists to study the structure and properties of seriously small particles.

DISCOVERY OF MICRO THINGS

Centuries ago, people thought mice came from grain since the creatures were always found running around grain barns. When the grain was gone, the mice seemed to vanish, too, so everyone thought mice and grain were connected in some way. Until 1665, this grain/mouse problem and many other phenomena (such as why do cheeses get moldy or why does meat spoil when left out in the open?) remained a mystery. Then, the first drawings of microorganisms were published in a book by Robert Hooke called *Micrographia*.

As a young man, Hooke, son of an English churchman on the Isle of Wight, was too interested in painting and building mechanical gadgets to spend time on his regular school homework. In fact, he didn't study much until he went to college at Oxford. There he spent long hours studying biology and math, trying to figure out how things worked. To learn more, Hooke needed to be able to see sample details clearly. At first he tried a magnifying glass, but when that wasn't good enough, he invented a compound microscope. With this tool, Hooke was the first to see tiny spaces inside a piece of cork. He called these spaces *cells*, like the small, single rooms of monks in a monastery.

In "Observation XVIII" of the *Micrographia*, Hooke wrote, "I could exceedingly plainly perceive it to be all perforated and porous, much like a honeycomb, but that all the pores of it were not regular…these pores, or cells…were indeed the first microscopical pores I ever saw." Hooke was also the first early scientist to draw the fine details of fleas, ocean sponges, and fossils.

Hooke's passion as an inventor and designer of scientific instruments also led him to build a wheel barometer, a spring balance wheel for watches, a universal joint in vehicles, and the first reflecting telescope.

Hooke's Law says that the amount a spring stretches is proportional to the amount of weight hanging from it.

Antony van Leeuwenhoek, a Dutch trader and son of a basket-maker, was also mechanically skilled. Hearing of Hooke's microscope and knowing that a microscope is only as good as its lens, Leeuwenhoek decided to build a microscope with a better lens. After grinding glass lenses, he built and installed them in a simple microscope that magnified samples up to 200 times. This allowed him to see the smallest hairs and details of his samples. But Leeuwenhoek wasn't great at art, so he hired someone to draw what he saw so that he could share his findings with others. Peering at a sample of pond water in 1674, Leeuwenhoek was the first to see the green algae *Spirogyra*.

Curious to see more, Leeuwenhoek examined samples of dental plaque collected from the teeth of his wife and children. Since his family sometimes brushed their teeth and tried to keep them clean, Leeuwenhoek decided he needed to do more testing, so he went looking for some "dentally challenged" subjects. After sampling the mouths of two old men ("who never cleaned their teeth in their whole lives"), Leeuwenhoek became the first person to see living bacteria under his microscope.

In a 1683 letter to the Royal Society of London, Leeuwenhoek described the bacteria as "an unbelievably great company of living animalcules, a-swimming more nimbly than any I had ever seen up to this time." Leeuwenhoek's biological discoveries, seen with an open mind and excellent instruments, were among the first and most accurate ever recorded.

Today we can see much more than Hooke, Leeuwenhoek, or any other of the early investigators were able to see. High-resolution microscopes and other instruments allow scientists to study the structure and properties of seriously small organisms and particles.

All About Scale

Remember that a nanometer is equal to a billionth of a meter—that's 10^{-9} meters. Think of a child looking at a mountain (about a factor of a thousand); an average 6-year-old child is 1/1000 the size of a mountain. An ant crawling around on the ground at the child's feet is about 1000 times smaller than the child. The ant has the same perspective of the child as the child has of the mountain. Bacteria on the body of the ant are 1000 times smaller than the ant, or about 1 millionth the size of the child. Finally, the sugar molecules powering the bacteria on the ant crawling around below the child is one billionth the size of the child.

So you see factors of a 1000 between sugar and bacteria, another 1000 to the ant, and another 1000 to the child. In other words, a nanometer is about $1/(1000 \times 1000 \times 1000)$ times smaller than we are. Yet scientists can control the structure and properties of such small items and create new things out of them!

STANDARDS

Observation and measurement are the keys to science and engineering. In research, as in other parts of life, we are constantly measuring using common units. The baseball cleared the fence by a foot. The NASCAR driver cruised at 185 miles per hour. The Olympic skier zipped into first place by 3/100 of a second. Standards of weight, height, volume, pressure, and temperature, for example, are important as references. It would be tough to measure anything without some known reference point.

A simple example of this is found in horse racing. It would be difficult to understand how a thoroughbred race horse could win the Kentucky Derby "by a length" if no one knew what a *length* was. In horse racing, a length is understood to be about the length of a horse, or about 2 meters. This rough measure works for horse racing because precise measure isn't needed. If the race is very close, though, a length won't do; instead, a photograph is taken at the finish line and the winner confirmed.

Science experiments and engineered devices must be far more accurate with strict standards of measurement. In mixing different chemicals or compounds together, exact units must be used or an erroneous end product may result. Similarly, for two metal gears to work together, they have to be machined to exact measurements. Close is not good enough.

> An *experiment* is a controlled testing of a sample's properties through carefully recorded observations and measurements.

Research is all about measuring and comparing. To repeat an experiment or follow someone else's method, a researcher must use the same exact units. It doesn't work to have a researcher in New York measuring in cups while in Germany samples are measured in milliliters, unless they can relate the two quantities to each other. To repeat an experiment and learn from it, scientists around the world need a common system.

International System of Units (SI)

In 1960, the General Conference on Weights and Measures adopted the *International System of Units* (or *SI*, after the French *Le Systeme International d'Unites*). When Great Britain formally adopted the metric system in 1965, the United States became the only major nation that didn't require metric measurements, though people had been using it since the mid-1800s. The *International Bureau of Weights and Standards* in Sevres, France, houses the official platinum standard by which all other standards are compared.

This early measurement standardization allowed scientists to compare "apples to apples." They could confidently compare data knowing that everyone understood. Internationally, this same standardization is needed for nanotechnology. Metric measurements are already in place, but industrial use and safety standards, exposure limits, and environmental release values must be determined alongside developing applications.

NOMENCLATURE

Nomenclature is the naming of things. In the scientific world, an international naming system exists for both chemistry and biology. In the chemical world, *chemical nomenclature* is used to communicate specifics about different chemicals and compounds.

> **Chemical nomenclature** is the standardized system used to name chemical compounds.

Chemical symbols, first based on Latin words, are used as an elemental code, because using the full name for a lot of chemicals can take a lot of paper or a lot of breath. The powerful insecticide dichlorodiphenyltrichloroethane—DDT—is written as $C_{14}H_9Cl_5$ using symbols to represent the total number of carbon, hydrogen, and chlorine atoms. Refer back to Table 1-2 for other examples of chemical nomenclature.

Chemical shorthand becomes especially important when writing down chemical reactions. The symbol for an element can be one letter, as in carbon (C) and potassium (K); two letters, as in silver (Ag) and copper (Cu); or three letters, as in recently discovered elements such as ununquadium (Uuq) and ununoctium (Uuo). Notice that when an element has more than a one-letter shorthand name, only the first letter of the symbol name is capitalized.

Periodic Table

In 1864, *Die Modernen Theorien der Chemie* (The Modern Theory of Chemistry) was published by German chemist Lothar Meyer. Meyer used the atomic weight of elements to arrange 28 elements into six families with similar chemical and physical characteristics. Sometimes elements seemed to skip a predicted weight. Where he had questions, Meyer left spaces for possible new elements. He also used the word *valence* to describe the number that equals the combining power of one element with the atoms of another element.

In 1870, Meyer's next generation of the Periodic Table with 57 elements was published. This table, including such properties as *melting point*, added depth to the understanding of interactions and the role of atomic weight. Meyer also studied the atomic volume of elements to fine tune his placement of elements into particular groups.

Perhaps Meyer's curiosity came from the fact that he grew up in a family of physicians and was exposed to scientific and medical discussions for much of his early life. His initial schooling in Switzerland was in the field of medicine. Meyer's chemistry research sprang from his fascination with the physiology of respiration. He was one of the first scientists to note that oxygen combined with hemoglobin in the blood. The many elements in the body and their complex interactions gave Meyer much to think about. To explain specific biochemical processes and systems, he needed to identify the elements more completely. Table 2-3 shows a few of the most common elements and their functions in the human body.

As of 2006, the Periodic Table contains 118 elements. Those up to atomic number 92 (uranium) are naturally occurring, whereas the "transuranic" elements, those synthesized in heavy nuclei interactions, make up the most recent discoveries. Some symbols of elements, as in Meyer's time, represent gaps or spaces for elements that have been hinted at by test data. When compared to Meyer's early table, the details described more than 150 years ago are amazingly accurate. The modern Periodic Table is shown in Figure 2-2.

Table 2-3 Elements perform a variety of functions in the body

Elements	Functions in the Body
Calcium	Bones, teeth, and body fluids
Phosphorus	Bones and teeth
Magnesium	Bone and body fluids, energy
Sodium	Cellular fluids, transmission of nerve impulses
Chloride	Dissolved salt in extracellular and stomach fluids
Potassium	Cellular fluids and transmission of nerve impulses
Sulfur	Amino acids and proteins
Iron	Blood hemoglobin, muscles, and stored in organs

Nomenclature is significant for regulators who have to identify and define nanomaterials. Some specific nanotechnology nomenclature challenges include nanomaterials of the same chemical composition but with different forms, such as carbon black, diamond, buckyballs, and nanotubes. Nanomaterials of the same chemical composition, but different size, such as *quantum dots*, will also need different definitions and regulatory standards. Quantum dots will be described in more detail in Chapter 8 when we discuss Smart Materials.

	IA	IIA	IIIB	IVB	VB	VIB	VIIB	VII			IB	IIB	IIIA	IVA	VA	VIA	VIIA	O
1	1 H																	2 He
2	3 Li	4 Be											5 B	6 C	7 N	8 O	9 F	10 Ne
3	11 Na	12 Mg											13 Al	14 Si	15 P	16 S	17 Cl	18 Ar
4	19 K	20 Ca	21 Sc	22 Ti	23 V	24 Cr	25 Mn	26 Fe	27 Co	28 Ni	29 Cu	30 Zn	31 Ga	32 Ge	33 As	34 Se	35 Br	36 Kr
5	37 Rb	38 Sr	39 Y	40 Zr	41 Nb	42 Mo	43 Tc	44 Ru	45 Rh	46 Pd	47 Ag	48 Cd	49 In	50 Sn	51 Sb	52 Te	53 I	54 Xe
6	55 Cs	56 Ba	57 *La	72 Hf	73 Ta	74 W	75 Re	76 Os	77 Ir	78 Pt	79 Au	80 Hg	81 Tl	82 Pb	83 Bi	84 Po	85 At	86 Rn
7	87 Fr	88 Ra	89 +Ac	104 Rf	105 Ha	106 Sg	107 Ns	108 Hs	109 Mt	110 110	111 111	112 112	113 113					

* Lanthanide Series	58 Ce	59 Pr	60 Nd	61 Pm	62 Sm	63 Eu	64 Gd	65 Tb	66 Dy	67 Ho	68 Er	69 Tm	70 Yb	71 Lu
+ Actinide Series	90 Th	91 Pa	92 U	93 Np	94 Pu	95 Am	96 Cm	97 Bk	98 Cf	99 Es	100 Fm	101 Md	102 No	103 Lr

Figure 2-2 Periodic table of elements

Biological Nomenclature

In the biological world, *zoological nomenclature* is used to communicate specifics about different genera and species. The *International Code of Zoological Nomenclature (ICZN)* is the organization that determines the scientific name of an organism. As in chemistry, biology uses specific naming standard rules that all scientists know and understand.

Researchers study microorganisms within an ICZN defined category. In fact, all living organisms are divided into separate groups and subgroups. When a plant or animal is named, it is categorized within a system of narrower and narrower characteristics (vertebrates versus invertebrates, for example). When it can't be defined further, the most basic level and scientific name has been reached (for example, *Homo sapien* is a modern human). The complete ICZN classification system has hundreds of branches describing in finer and finer detail the differences between species.

Nanotechnology Nomenclature

As more and more new materials are developed at the nanoscale, nomenclature is becoming increasingly important. Everything from C_{60}, C_{70}, and C_{80} to SWNTs, quantum dots, and gold nanoshells are now part of the biological and materials research forefront. A common nomenclature will make it much easier for everyone from scientists to policy makers to the public to use and understand these materials and concepts.

Professor Vicki Colvin of the Center for Biological and Environmental Nanotechnology at Rice University is working toward establishing standard nomenclature for the entire nanotechnology industry. She presented her nano-nomenclature ideas in 2005 at the American Chemical Society's 229th national meeting in San Diego, California. This forward-thinking move will make interdisciplinary research much easier.

Colvin notes that the number of "nano papers" published in all the scientific literature increased from 0 in 1990 to nearly 20,000 in 2005, reflecting the growing importance of the subject. Depending on a nanoparticle's formation and use, it can be called by lots of different names. Colvin suggests that a system similar to that used for polymers could be developed for nonotechnology. She suggests that a particle's surface chemistry properties are a good start toward defining nanoparticles.

Risk assessment needs a comprehensive naming system to avoid misunderstandings that could be critical. Environment cleanup also needs an "apples to apples" dictionary of terms to use for the development of good air, water, and soil regulations, as well as worker safety standards.

Since the nano world is at the foundation of matter (atoms and molecules), it naturally cuts across nearly every major field of research. This is one of the most important features of nanoscience and nanotechnology—its amazing flexibility touches everything! Every industry will feel its broad influence in some way.

Quiz

1. Leeuwenhoek saw microorganisms in
    (a) polio sufferers
    (b) belly button fuzz
    (c) malaria victims
    (d) dental plaque

2. One of a nanoparticle's most important properties is much greater
    (a) smell
    (b) acidity
    (c) surface area
    (d) cost

3. Today, with high-resolution and electron microscopes, scientists can see
    (a) much less than early scientists were able to see
    (b) much more than early scientists were able to see
    (c) about the same as early scientists
    (d) nothing at all

4. When a spring stretches in proportion to the amount of a suspended weight, it is known as
    (a) Joel's Law
    (b) Hooke's Law
    (c) Douglas's Law
    (d) Crooke's Law

5. The chemical nomenclature $C_{14}H_9Cl_5$ is short for

   (a) MTV

   (b) ADD

   (c) DDT

   (d) STM

6. In the biological world, what standard naming method is used?

   (a) Periodic Table

   (b) International Bureau of Weights and Standards

   (c) Standard Isotopes and Radiation Chart

   (d) International Code of Zoological Nomenclature

7. The main difference between using nanotools and regular lab equipment is

   (a) size and scale

   (b) smell and sound

   (c) length and breath

   (d) freezing and thawing

8. A controlled testing of a sample's properties through carefully recorded observations and measurements is called a(n)

   (a) experiment

   (b) laboratory

   (c) science fair project

   (d) hypothesis

9. A nanometer is equal to a

   (a) zillionth of a meter

   (b) millionth of a meter

   (c) billionth of a meter

   (d) trillionth of a meter

10. What types of tools do nanotechnology technicians and scientists use?

   (a) tiny hammers and wrenches

   (b) global positioning systems

   (c) compasses and light

   (d) optics, magnetics, and electricity

CHAPTER 3

What Makes Nano Special?

So what's the big deal about nanotechnology? The simple answer is that super small things like atoms behave differently than the same stuff behaves when it's bigger. Individually, the basic building blocks of matter don't have much of an effect on their environment. Everything is too small to have much substance or mass. But when they combine into larger forms, unique properties and functions emerge.

We now have the ability to tailor matter in important ways. Nano exploits the *quantum* properties of matter (more in Chapter 10), which is not the same as saying atoms don't have an effect on their environment, because they do; it's just that their environment is much, much smaller than our macroscopic perspective.

Quantum is the smallest physical amount that can exist independently such as a separate quantity of electromagnetic radiation (energy).

In 1959 "There's Plenty of Room at the Bottom," Richard Feynman said, "The principles of physics, as far as I can see, do not speak against the possibility of moving things atom by atom. It is not an attempt to violate any laws; it is something, in principle, that can be done; but in practice, it has not been done because we are too big."

Size controls the nanotechnology equation.

To understand this concept better, imagine yourself sitting on the 50-yard line in the middle of a huge football stadium that holds around 70,000 people. You are alone. If you shouted at the top of your lungs, would you be heard in the upper decks? If you ran up and down the aisles or bleachers, would anything in the stadium be affected? Would your body heat impact your surroundings? If you stamp your feet, would the vibration be carried into the next section? The answers, of course, are all no. In a large space, one solo individual doesn't physically impact his or her environment much.

Now imagine the same stadium filled with thousands of excited, screaming, die-hard football fans. Do they impact their environment? Yes! Noise levels are probably beyond OSHA-safe limits, as fans stomp and clap their approval of successful plays and touchdowns, sending out vibrations that can be felt in the rafters. The air conditioning is running full out to keep up with the sweating, waving masses. Food concession stands are crammed, and bathroom waiting lines are long.

In the nano world, particles can be close together or far apart, but because of special, size-dependent properties, interesting things happen. As you learned in Chapter 2, size affects nanomaterial properties in different ways (such as color, surface area, conductivity, and strength). Gravity and Newton's laws also affect nanoparticles in odd ways. Gravity influences mass, and nanoparticles have hardly any mass. Therefore, gravity does not affect nanoparticles to any great extent.

When Feynman said there is a lot of room at the bottom, he was referring to the huge opportunity that the nano world offers for new research, but the nano world is also the hangout of atoms and molecules. Since atoms and subatomic particles are miniscule, there is a lot of room between them as well. However, very few isolated atoms hang out on Earth. Most elements (except the noble gases) exist in molecular or compound form.

Atoms are in everything; they interact with everything. We eat them, breathe them, clothe ourselves with them, and get them on us constantly. Everything, absolutely *everything* we know about is made from atoms or parts of atoms.

Carbon Forms

In previous chapters, you learned about the different structures that carbon can take and how bonding affects function and strength. Since a lot of exciting nanotechnology re-

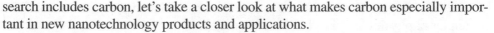

search includes carbon, let's take a closer look at what makes carbon especially important in new nanotechnology products and applications.

Carbon, the sixth most abundant element in the universe, has been familiar to humans since ancient times. Carbon is found in the atmosphere and dissolved naturally in water. In rock, carbon forms carbonates of calcium (limestone), magnesium, and iron. It has been dug from coal deposits and processed into useful forms like heating fuels for centuries. Nearly 10 million compounds contain carbon (that we know of), and hundreds of thousands of those are critical to organic and biological processes.

Three naturally occurring carbon *allotropes* (forms) exist as amorphous, graphite, and diamond. (Buckyballs and nanotubes, while initially created in the laboratory, are now known to form naturally in soot from flames.) *Amorphous* carbon is created when a carbon-containing material burns without enough oxygen to burn completely. This black soot (also known as lampblack, gas black, or carbon black) is used in paints, inks, and rubber products. It can also be compressed into different shapes and used as the cores of dry cell batteries.

Graphite, one of the softest carbons known, is commonly used as a lubricant. Although found naturally, commercial graphite is usually made by treating petroleum coke (black tar deposit left over from crude oil refinement) in an oxygen-free oven. Naturally occurring graphite is found in two forms. Although they have identical physical properties, they have different *crystalline structures*. All artificially made graphite is the *alpha* type. In addition to being used as a lubricant, a lot of graphite (in a form called *coke*) is used in steel production. Coke is made by heating soft coal in a furnace without any oxygen mixing with it.

Diamond, as you learned in Chapter 1, is one of the hardest substances known and the third naturally occurring carbon form. Natural diamonds are usually used for jewelry, while commercial quality diamonds are often artificially created. Small industrial diamonds are made by squeezing graphite under high temperature and pressure for several days or weeks. They are used for such applications as diamond-tipped saw blades.

When large carbon-only molecules (Buckminster fullerenes) were discovered, everyone sat up and took notice. Researchers tried to decipher how carbon could wear so many disguises (structurally) and still basically be carbon. This single molecule with 60 or 70 carbon atoms (C_{60} or C_{70}) linked together like a soccer ball was an amazing discovery. Further research discovered how these carbon molecules trapped other atoms within their round cages. These carbon types had very different physical properties (and crystalline structures), but they are able to withstand great pressures, and they possess magnetic and superconductive properties at high temperatures. This great versatility was not available to their better known cousins, graphite and diamond.

Single Walled Carbon Nanotubes

Since Galileo's time, new devices and precise instrumentation in the hands of the well-trained observers have hastened the discovery of new insights into nature. The development of new instrumentation was essential to this process. In that light, nanotechnology has also been a result of applied or directed research.

You'll remember that Richard Smalley invented an instrument that discovered fullerenes and turned chemistry on its ear. While the chemistry of fullerenes is becoming well-established among researchers, the chemistry of single-walled carbon nanotubes (SWNTs), commonly called *buckytubes,* has been a hot area for years. In fact, many researchers think the application potential for nanotubes far outstrips that of buckyballs. Investigations into SWNT bonding characteristics are important because they provide basic structural insight into how nanoscale interactions take place. Understanding SWNT chemistry, then, becomes critical to well thought out, predictive management of their properties. Some methods include forming metal complexes with SWNTs to control site-selective chemistry during fabrication.

Smalley's team was hot on the trail of finding a way to synthesize infinitely long, symmetrical SWNTs. He believed SWNTs had almost unlimited potential to change traditional energy transmission methods. He found that buckytubes could be made to conduct electricity very efficiently at a fraction of the weight of metal wires without sacrificing material strength. In fact, there is the possibility of placing a "tube within a tube" that could form a wire with an insulating outer shell and a conducting inner shell. Since a buckytube's properties are size-dependent, diameter and other dimensional changes can be fine-tuned. Unlike bulk materials, the properties are not chemically related but are connected to physical geometry. Figure 3-1 illustrates how a tube within a tube compares to current copper wiring.

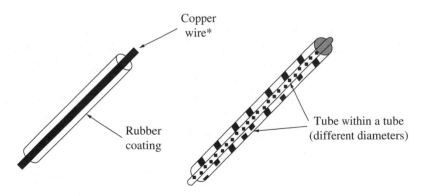

Figure 3-1   Concentric SWNTs could serve as super efficient electrical transmission wires.

Nanorods

Nanorods are made from silicon, metals (such as titanium, tin, and zinc), other semiconductors, and insulators. Depending on the starting material, nanorods have different electronic, optical, and mechanical properties that work in a variety of applications, including electronics, sensors, optical components and displays, polymer composites, and actuator devices.

Creation of nanorods is usually done in a vapor phase (for example, evaporation/condensation) or by wet chemistry (chemistry that occurs in water) methods. However, they can also be grown on gel substrates with electrophoretic deposition (that is, by drawing of a mixture of materials through a fine gel with an electric current) and high heat for crystallization.

After nanorods are created, they can then be self-assembled into larger nanomaterials and structures with highly reactive surfaces.

Color

So how can carbon not be carbon? Or gold not golden? How can an element be the same element, yet different?

Once again, size is the determining factor. A gold ring on your finger is gold-colored, but if you shrink that gold down to its nanometer size (10 nm to 100 nm), it's red. You can realize different optical properties as a result of a material's reduced size. That doesn't work in the "big world"—that is, you can't change a bulk material's color just by slicing it up into smaller pieces.

With nanoparticles, however, it's a whole new ballgame. *Nano-color* is a physical characteristic. An element's nanoscale composition may not be what you expect!

Medieval and Victorian glass workers stumbled upon the color changing property of gold. They worked the metal in their forges until they produced nanoscale gold particles that were green, orange, red, or purple depending on their size. A lot of the colored glass in the stained glass windows of churches of the time was a result of their chemical composition as well as beautiful artistry. Some early glazes used in ancient pottery also got their colors from these special nano-color changes. The early potters were good at producing beautiful multi-colored finishes, but they didn't understand the nanoscale properties that caused these colors.

Surface Area

When you start making things nanoscopic, you start to get large numbers of them (we're talking billions on the head of a pin). Imagine a chunk of metal that you cut up into smaller and smaller chunks until you get down to nano-sized bits. You've got billions of nano-sized bits, but when added together they make up the same amount of metal material that you started with.

In fact, you may be able to make 10^{15} bits out of one big chunk of metal, but you actually end up with much more reactive area. Why? Because of surface area. Picture the original metal chunk with four sides. As the first chunk is divided into ever smaller parts, the surface area is also increasing. If you added up the surface area of each chunk every time you cut it up, you'd get a huge amount of surface area. The surface to volume ratio gets very large. Figure 3-2 illustrates special color and surface area properties at the nanoscale.

To put it into perspective, think of a basketball that's made of solid metal. Slice it up into nano-sized particles until you get about 10^{15} particles. The surface area of a 1 m diameter basketball is around 2 square meters. The surface area of the numbers of nanoparticles making up the same amount of material would equal about 10,000 football fields!

As you've seen, scale is important in nanotechnology. A length scale that spans 10 angstroms (Å) to 100 Å could include 100 to 1000 particles. This super small size makes surface effects so important. Table 3-1 lists the sizes by which nanoparticles can be divided.

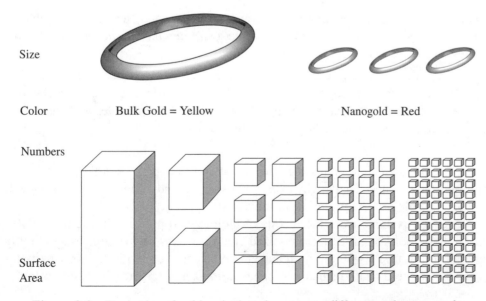

Size

Color   Bulk Gold = Yellow   Nanogold = Red

Numbers

Surface
Area

Figure 3-2   Properties of gold and other elements are different at the nanoscale.

Table 3-1 Nanomaterials are categorized according to their size dimensions

Nanoparticle Dimension	Nanomaterial
All three dimensions < 100 nm	Particles, quantum dots, nanoshells, microcapsules, hollow spheres
Two dimensions < 100 nm	Single-walled carbon nanotubes, fibers, nanowires
One dimension < 100 nm	Anti-adhesive/anti-stain coatings, applied films, viruses

Now think of a hamburger. Remember all the media attention about safe handling of hamburger? The original meat is sliced and diced until its surface area is much greater than the original steak's surface. Add bacteria and improper handling techniques to this vast surface area and it can lead to botulism (courtesy of *Clostridium botulinum* bacteria) or other toxins. You eat it, you die (without treatment, that is; death is caused by suffocation from neurotoxin effects).

So when researchers make things such as detectors or catalysts with nanomaterials, they get materials that are much, much more efficient because there's so much more surface area to work with. It's another major benefit that makes the nanoscale special.

Quantum Mechanics

Continuum mechanics that direct things such as gravity and Newton's laws change when you drop down to the nanometer scale. *Quantum mechanics* or discrete processes take over. They get really, really strange.

> **Quantum mechanics** is the physics theory that describes the special properties of matter at the smallest levels. Many nanoparticles and nanostructures behave according to these theoretical rules.

Again, scientists find nanoscale effects that don't happen when they're working at larger bulk scales. That's why everybody's so excited about nanotechnology. It creates a whole new scientific opportunity.

Here's an example from the worlds of biology and physics. In the mid-1990s, researchers took round, nanometer-sized silica particles (same stuff as sand, only smaller) of around 20–30 nm in diameter and coated them with very tiny particles of gold. The gold-coated silica was called a *nanoshell* (created by Rice professor

Naomi Halas—more in Chapter 10) and was something like a super small candy-coated chocolate.

The nanoshell core size and gold thickness could be made thicker or thinner. Researchers found that when the nanoshells reflected light, their color changed as their size changed. Reflectance would be red, purple, or green depending on their size and shell thickness.

If the nanoshell was made just right, nano gold would absorb infrared light. This was another important nanoscale finding since infrared light passes through tissue up to about 5 inches deep. This opened the door to lots of medical possibilities in several different areas. (We will take a much closer look at the great medical benefits of nanotechnology in Chapter 6.)

Manufacturing

Nanotechnology allows for new methods of manufacturing. At the beginning of the twentieth century, in order to build circuits and materials that were seriously small, everything had to be done with painstaking precision and care in super clean rooms. Even then, the end result was sometimes not perfect.

Engineers and scientists were always on the outside looking in, trying to build machines that were smaller and faster, and had more memory. This type of manufacturing, called *top down,* involves large starting materials that are shaped and formed into smaller and smaller devices. Think of a sculptor starting out with a huge chunk of marble and forming it into something of intricate beauty. Michelangelo used to say that he chose a stone and then proceeded to release the figure within. Nanotechnology does something similar in manufacturing. Scientists start with atoms that were there all along (as the basic forms of matter), and move them into useful combinations. Today, nanotechnology allows materials to be created from scratch—that is, things are constructed molecule-by-molecule or even atom-by-atom.

BOTTOM-UP VS. TOP-DOWN NANOTECHNOLOGY

Self-assembly of microcapsules and other nanoparticles is done from the *bottom up.* Collecting, combining, and shaping atoms and molecules into specific structures by chemical and catalytic reactions are how engineers and scientists hope to fabricate components at the nanoscale level. This is how computer and electronics designers hope to build computer chips so small they fit on a fingernail or could be implanted under the skin. You'd always be able to phone home!

Top-down manufacturing is how computers are made today. Smaller and smaller components are etched with delicate instruments, chemicals, and a template onto a layered substrate to get the right circuitry.

The type of manufacturing depends on the starting materials. Although nanoparticles have not been formally classified and standards established (more about this in Chapter 13), general categories can be described.

Products

Nanotechnology isn't all in the future. Manufacturers are enhancing old products with new nanoproperties that go beyond those available in bulk material. By manipulating materials at the molecular level, scientists can improve product durability and strength.

You probably didn't even realize that lots of products on store shelves today come from nanotechnology or contain nanoparticles. For example, strong sunscreen is no longer the bright white stuff you put on your nose that makes you look like an alien. Now it's transparent and you can spray it on. That's nano-sized titanium dioxide and zinc oxide, chemicals whose high surface area provides a lot more deflection of the sun's harmful UV rays. It works just as well as larger white particles in earlier products, but it doesn't scatter visible light and doesn't have the blazing white color.

Other examples of nanomaterials include nanoscale silica used as dental fillers and nanowhiskers used in stain-resistant fabrics, such as Eddie Bauer's Nanopants. Nano-Tex, the company that developed Nanopants fabric technology, makes coatings that stick to fabrics at the submicron (micrometer) level. This lowers the amount of chemicals needed to clean clothes and transfers stain resistance onto each individual fiber. More than 80 textile mills around the world use Nano-Tex technology in products sold by major clothing and furniture brands.

The great benefit of nano-treated fabrics is that the fabric becomes *hydrophobic*, or repellent to penetrating stains. So instead of the old top-down method of trying to get the stain out after it happens, the stain is prevented from sticking to fibers in the first place (bottom-up prevention). It's possible that nano-coatings could one day make clothes washing machines obsolete.

Another new household product keeps your home's surfaces looking as neat and clean as your clothes. It uses nano titanium dioxide, like the sunscreen, and catalyzes the breakdown of organic materials. In other words, it makes dirt slide off windows and other surfaces. Windows, windshields, or even reading glasses treated with new nanotechnology anti-adhesive products will be self-cleaning, which means less back-breaking work and more time for golf, garden club, and computer games!

Nanoscale engineering of surfaces and layers leads to products with better mechanical, wetting, thermal, biological, electronic, optical, and chemical properties. Some of these are listed here:

- Protection against wear for machinery and equipment
- Protection of soft materials (for example, polymers, wood, and textiles)
- Anti-graffiti and anti-fouling coatings
- Self-cleaning surface films for textiles and ceramics
- Corrosion protection for machinery and equipment
- Heat resistance for turbines and engines
- Thermal insulation for equipment and building materials
- Biocompatible implants
- Anti-bacterial medical wound dressings and tools
- Super thin components for transistors
- Magneto-resistant sensors and data memory
- Photochromic and electrochromic windows
- Anti-reflection screens
- Smaller and more efficient solar cells

Nanobots

It's easy to imagine all kinds of possibilities for manufactured materials when you start at the atomic level; however, some projects are either impossible or unlikely to be developed. One of these projects is the imagined development of nanoscale robots (*nanobots*) working on their own with ill intentions. Nanobots (also known as molecular assemblers) are the villains of fictional scenarios—the "gray goo" destruction of the universe's inhabitants—created by nanotechnology pioneer K. Eric Drexler in his book *Engines of Creation*. (In a June 2004 edition of *Nature* Drexler said he wished he had never mentioned gray goo!) Other shocking imaginations by fiction writers such as Michael Crichton in *Prey* and writers of Saturday morning cartoons involve superheroes who fight nanotechnology weapons of mass destruction let loose by alien bad guys to devour the Earth or universe.

Add this to media hype, and it equals a sandstorm of bad information that has been unleashed about what might happen centuries in the future—versus what could never happen because of the laws of physics and entropy. Like most science fiction, the science tends to get pushed out by the fiction.

When it comes to these proposed nanotechnology horror stories, researchers agree that nanoparticles' extremely miniscule size keeps them from moving fast or in a directed way. They also don't have the power source needed to function in ways used by living biological organisms. It is even possible that nanobots made from organic molecules could find themselves being preyed upon by common bacteria or fungi. If fictional nanobots could reproduce from inorganic materials (such as rocks) into workable components, then synthesis itself would take all their time, attention, and resources.

Computer scientists describe how fictional nanobot programming would have to overcome huge hurdles (even if they ever got past the whole energy synthesis problem). Programming safeguards would include everything from instructions to power down after a certain period of time, to needing constant updates or rare materials to thrive. It's fun to contemplate these "technology gone wild" story lines, but the reality is that nanobots and their nefarious capabilities are fantasy.

Serious nanotechnology researchers such as the late Richard Smalley and professor of chemistry Mark Ratner (expert in molecular electronics at Northwestern University and recipient of the 2001 Feynman Prize in Nanotechnology) discredit nanobots as fun to read about for entertainment, but lacking in real science substance and viability. Smalley cited the massive amount of time it would take to build a bulk material atom-by-atom. Remember that compounds are made up of zillions (10^n) of atoms—in other words, an unimaginable number of atoms. One example compares an atom's volume to a teaspoonful of water. To make a few ounces or milliliters of water atom-by-atom would be like filling the Pacific Ocean a teaspoonful at a time.

Quiz

1. When a nanoshell's core size and gold thickness are changed, it changes the
   - (a) usefulness
   - (b) color
   - (c) smell
   - (d) texture

2. Lampblack, gas black, and carbon black are all names for
   - (a) lead
   - (b) nitrogen
   - (c) mercury
   - (d) amorphous carbon

3. Medieval glass workers stumbled upon the color changing property of

   (a) tin

   (b) potassium

   (c) gold

   (d) iron

4. Self-assembly of microcapsules and other nanoparticles is done from the

   (a) top down

   (b) bottom up

   (c) lateral view

   (d) z direction

5. Nanotechnology allows materials to be created from

   (a) huge blocks of granite

   (b) atoms and molecules

   (c) dust bunnies

   (d) meteors

6. Which nanomaterials are made from silicon, metals, semiconductors, and insulators?

   (a) nanorods

   (b) sculptures

   (c) nano limbs

   (d) hot rods

7. Carbon is how abundant in the universe?

   (a) second most abundant element

   (b) fourth most abundant element

   (c) sixth most abundant element

   (d) rarely found

8. Nanobots are all of the following except

   (a) fictional

   (b) a self-admitted mistake by Eric Drexler

   (c) nearly impossible to manufacture, program, and power

   (d) a real, looming threat to humanity

9. Silica particles of around 20–30 nm in diameter and coated with gold are called

    (a) nanoballs

    (b) golden globes

    (c) nanoshells

    (d) nanogels

10. When large starting materials are formed into smaller and smaller parts, it is called

    (a) top-down manufacturing

    (b) a time-consuming process

    (c) bottom-up manufacturing

    (d) corporate profits

CHAPTER 4

Nanoscience Tools

In *Star Trek* and *Star Wars* films, we've met beings as different from one another as Wookiees and Klingons. Today's space telescopes show us places where only our imaginations have traveled. Alien stars and landscapes, so long beyond our reach, are now studied and analyzed for everything from chemical composition to cloud cover.

Discoveries are not just happening "out there" but here on Earth, too. Technology is charging ahead so fast that we can now see things that are seriously small. Strange and hidden worlds only imagined in the past are coming into sharp focus through scientific discoveries and engineering advances.

Nanotechnology, the new and exciting field currently pursued by many scientists, engineers, computer designers, and business people, didn't arrive overnight. Like many new sensations, many years of thought and research and the invention of new tools paved the way for a boatload of new technology.

Tools of Discovery

Although early scientists first spotted living bacteria with a light microscope using a finely ground lens, today's researchers use high-powered and complex instruments to study nanomaterials and nanostructures. The human eye can tell the difference between two objects that are 0.1 mm apart at 25 cm distance; any smaller detail must be magnified to be seen.

Regular light microscopes are not nearly powerful enough to see molecules. Light microscopes have a limit of about $1000\times$ (1000 times actual size) magnification, and the smallest objects you can see separately with white light are about 200 nm in length. Researchers in the emerging field most often use electrons, not light, to study the secrets of molecules, compounds, and nanomaterials. They examine samples at high magnification using electron microscopes.

> *Magnification* is the amount an image is enlarged in a microscope.

Whether the samples *fluoresce* (emit light at a certain wavelength) when activated by light of another wavelength, contain metal ions, or come from a biological source makes a big difference as to which instrument is used to view them.

ELECTRON MICROSCOPES

Scientists use several types of electron microscopes to study the various properties of nanomolecules:

- Scanning electron microscope (SEM)
- Transmission electron microscope (TEM)
- Analytical electron microscope (AEM)

Depending on the sample and what information is needed, these and other related instruments are used to study everything at the nanometer scale.

> *Electron microscopes* use the energy from electrons to magnify samples from 10 to 1,000,000 times.

Scanning Electron Microscope

The *scanning electron microscope* (SEM) uses a focused beam of electrons to scan the surface of thick or thin samples. The SEM makes an image appear a lot like a stereo microscope or dissecting microscope image in biology—it has a 3-D appearance. But in the SEM, images are produced one spot at a time in a grid-like pattern.

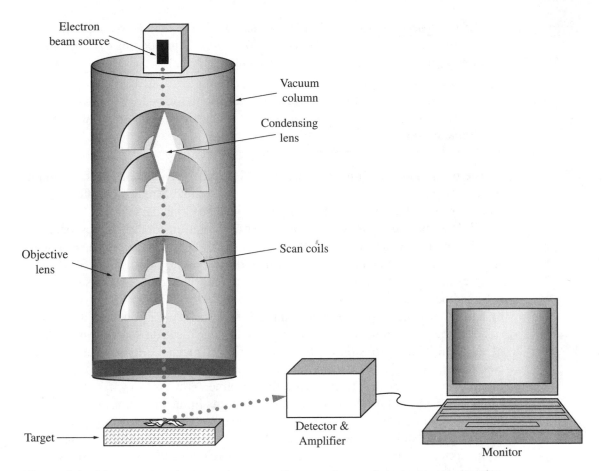

Figure 4-1 The scanning electron microscope focuses a beam of electrons on a sample.

Figure 4-1 illustrates the main parts of a scanning electron microscope.

SEMs provide good resolution of a variety of samples, typically 5 to 10 nm, or magnifications of 10 to 100,000× ; some of the newest SEMs can reach 1,000,000× magnification, with a resolution of 1 nm.

Resolution is the amount of clear detail that is visible in an image.

You can enlarge a photograph a lot using extremely powerful lenses, but the image will be blurry and the quality will be reduced the larger you get. Increasing the magnification doesn't increase the resolution.

The *resolving power* of an optical system is limited by the *wavelength* involved, and *diffraction* (amount light is bent) by the *aperture* (lens opening). A bigger aperture has a greater resolving power than a smaller one.

> **Resolving power** is the measure of the ability of a lens or optical system to form separate and distinct images of two objects with small angular separation.

The black-and-white images created on SEM video screens match the scanning on the screen to the scanning of the beam on the sample. Areas that reflect a lot of electrons look bright white, while nonreflective areas look dark.

SEM samples are normally studied under a vacuum inside the microscope; they must be prepared carefully so that they don't shrink or change shape. Biological specimens are dried and coated so that they don't shrivel. Because the SEM views samples with electrons, they also have to conduct electricity. SEM samples are mounted on a platform called a *target peg* and coated with a very thin layer of metal by a machine called a *sputter coater;* this gives the sample the conductivity needed.

After the air is pumped out of the microscope, an electron source sends a beam of high energy electrons down through a series of magnetic lenses that focus the electrons to a very fine point. A set of scanning coils (see Figure 4-1) moves the focused beam back and forth across the specimen, row by row. As the electron beam hits each spot on the sample, other electrons in the atoms of the sample and the coating are knocked loose from the surface. A detector counts these electrons and sends the signals to an amplifier. The final image is then collected point-by-point from the number of electrons emitted from each sample spot and shown on a computer screen.

Transmission Electron Microscope

While SEMs can only skim a sample's surface, *transmission electron microscopes* (TEMs) see all the way through a sample. A wide beam of electrons passes through a thinly sliced sample to form an image. The electrons that a TEM uses to project images are bent by magnets, like a glass lens bends light in a regular microscope. In this way, electrons can be focused until a clear picture appears on the monitor.

This microscope is similar to a regular upright light microscope, in that thicker or denser areas of a sample absorb or scatter the beam, causing dark areas, while thin or less dense areas look light. Since most biological samples are made of carbon, nitrogen, oxygen, and hydrogen, not enough density difference exists in the structures to see dark or light in a TEM. Biologists sometimes stain a sample by chemically attaching dense metal atoms to specific atoms or molecules in a sample to make them appear darker in an image.

A TEM can see images 1000 times smaller than a compound microscope and about 500,000 times smaller than a human eye. The resolution of a TEM is about 0.1 to 0.2 nm. This is the typical separation between two atoms in a solid.

Electron microscopes use a high energy electron beam to see very small (nanoscale) samples and determine specific properties such as the following:

- Size and shape
- Texture and complexity (how a sample looks)
- Makeup (amounts of elements, compounds)
- Structure (arrangement of atoms and molecules)
- Properties (melting point, hardness, strength, conductivity, reactivity)

Today we see much more than Hooke, Leeuwenhoek, or any of the other early investigators were able to see. We can see within cells, substructures, chromosomes, proteins, and individual molecules and atoms.

More Power—Analytical Electron Microscope

The ability to measure material structure and chemical characteristics atom-by-atom is an important key to future progress in nanoscience, nanomaterials, and nanotechnology. Transmission electron microscopes are built to look inside materials at high magnification. As a multipurpose tool for materials discovery and characterization, it is unmatched by most other methods

A TEM equipped with analytical instruments such as X-ray and electron spectrometers is called an *analytical electron microscope* (AEM). These machines measure and form images from the X-rays created by atoms when they are bombarded by lots of electrons, and they can also measure the loss of energy that electrons suffer when they pass through materials. These measurements can see the differences between carbon atoms and nitrogen atoms and between iron and nickel atoms, for example, allowing scientists to map out the composition of a material in fine detail.

The AEM's ultra-high performance provides scientists with extremely high-resolution imaging capabilities (up to 0.1 nm). The AEM can also provide information on a material's atomic composition, molecular bonding, and electrical conductivity. With the AEM, researchers can get detailed information on nanomaterials' molecular makeup as well as data on specific properties and performance of things made from it.

An AEM permits detailed study of advanced materials for technological applications. Some areas to benefit from the AEM scanning include biomedical research, smart coatings, fuel cell research, magnetic nanostructures, and semiconductor quantum dots. These applications require in-depth atomic structural data on interfaces, boundaries, and possible faults.

SCANNING PROBE MICROSCOPES

Scanning probe microscopes (SPMs) are a group of instruments used to study the surface character of materials from the atomic to the nanoscale. These microscopes have a probe tip, or *stylus*, either fixed or mounted on the end of a tiny beam, that tracks and records sample surface changes as the surface of the sample is moved in a grid pattern. The tip tracks along a sample's surface like a Mars land rover, recording the height, electrical, or other changes on the surface. A SPM works a lot like the old phonographs that played records with a "needle" tip that tracked and vibrated up and down and side to side in grooves as the record went around on the player.

The up and down movements of the tip of the beam are measured by a laser beam that reflects off the top of the beam, and its jiggling as the beam vibrates is measured by a optical detector that produces an image of the surface. The voltage difference between the sample and the tip can also be measured, as can the amount of current flowing from one to the other. The sample is moved in a scanning mode by *piezoelectric crystals*.

> Crystals that have the ability to generate a voltage in response to applied mechanical stress are known as *piezoelectric crystals*.

Scanning Tunneling Microscope

The first SPM was the *scanning tunneling microscope* (STM), invented in 1981 by Gerd Binnig and Heinrich Rohrer at IBM's lab in Zurich, Switzerland. This revolutionary microscope uses a fixed probe tip to measure a material's surface electrical characteristics and won the Nobel Prize for physics for the scientists in 1986.

Various kinds of images can be seen, depending on the type of STM probe tip used. The simplest method is to scan the structure of a surface, keeping the tip a constant distance away, such as one atomic diameter or 0.2 nm, and the tip is raised or lowered to keep the current flowing at a constant value, which means the distance is maintained. Usually a voltage is created between the probe tip and the conductive sample's surface to cause a flow of electrons across the gap (tunneling). Figure 4-2 illustrates a STM tungsten tip.

> *Tunneling current* is created when a sharp conducting tip is placed very close to a conducting surface with a voltage difference between them. Electrons then flow across the gap in proportion to the gap distance.

When the tip moves up or down relative to the surface to keep the tunneling current regulated, it also tracks and records the sample's topography (surface). The

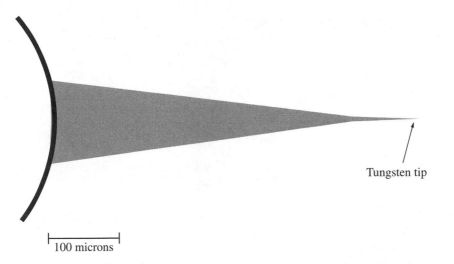

Tungsten tip

100 microns

Figure 4-2 A tungsten tip is often used for a scanning tunneling microscope.

amount of current is dependent on the distance between the probe and the surface. For current to be maintained, the sample must be conductive. Insulating samples that are not conductive, such as rubber, can't be scanned by STM unless they are first coated with a metal or another conductive coating.

The tip is very sharp and the end is only one atom wide. This means that the STM makes it easier for scientists to see and position individual atoms with high resolution. It can be used to capture images from conductive surfaces at about 0.2 nm resolution. It can also cause chemical reactions and create ions by pulling off individual electrons from atoms and then converting them back by adding the electrons.

Atomic Force Microscope

The *atomic force microscope* (AFM) uses a laser that reflects off the back side of the probe tip to find the position of the probe tip, which is mounted on a flexible beam. (Remember this is too tiny to see with the naked eye!) Ridges and valleys on the sample surface cause the probe tip to lift and drop as the tip is dragged across the surface. The AFM can operate in several modes, such as in the contact mode just described, where atoms repel each other in close contact, or in non-contact mode, where the probe is held just above the surface and the small attractive forces between atoms not in contact are measured.

An AFM has two measures of resolution: the plane of the measurement and the direction perpendicular to the surface. The sharper the probe's tip (usually made of silicon or silicon nitride), the better the resolution, since it can get into smaller spaces. A super sharp carbon nanotube tip is the ultimate AFM tip and can follow a

Figure 4-3 A single walled carbon nanotube probe tip increases accuracy.

Single-walled carbon nanotube

sample's contours very accurately. Figure 4-3 illustrates how a carbon nanotube tip fits into spaces in a sample's topography. Figure 4-4 illustrates how better resolution is created through finer probe tips.

AFM vertical resolution is obtained by comparing relative vibrations of the probe above the surface. Extraneous vibration sources include sounds, floor/building vibrations, and thermal vibrations. Getting the best vertical resolution requires stop-

Tungsten probe tip

Carbon nanotube probe tip

Figure 4-4 In atomic force microscopy, the finer the probe the better it traces the sample.

ping instrument jiggle (a non-technical term). Many instruments are placed on vibration-free counters or hung from bungee cords that dampen extra jiggling.

Multifunctional Microscopes

Since the mid-1800s, the resolution limit due to diffraction (scattering) of light in small apertures (holes or edges of particles) has limited optical image resolution to the wavelength of light (0.5 micron or 500 nm for visible light). Development of SPMs in the 1980s gave imaging three orders of magnitude better resolution. However, basic physics limits measurements in all types of microscopy.

STM images are defined by electron tunneling physics, while AFM has a wider range of capabilities. It can respond to a broad range of forces between tip and sample/surface, such as friction, magnetic, Coulombic (like charges repel each other), atom core repulsion, and chemical attraction (desire to bond).

Optical imaging can augment STM/AFM images. Although light diffraction or scattering limits us in ordinary microscopes, *near-field scanning optical microscopy* (NSOM), a type of microscopy where a sub-wavelength light source is used as a scanning probe, gets around that limit by placing the light up very close to a sample (a few nanometers above the surface). This can be done by placing a very tiny hole near the sample (closer than half the hole diameter, or about 50 nm) and shining a strong light source (a laser) at the hole. Enough light gets through the hole to illuminate the sample, and a detector can record the light reflecting off the sample. Several NSOM methods have been worked out that use tiny holes or tips of glass optical fibers with visible light image resolution of around 10 nm.

NSOMs are not limited to visible wavelengths. Work can also be done that allows several operations such as detection, excitation, reflection, and transmission to be performed. It can be used to study and test everything from electronics to electrical signals between cells.

Laser Scanning Confocal Microscope

A *laser scanning confocal microscope* (LSM) uses a laser (ultraviolet) light and scanning mirrors to sweep across a sample. A small aperture limits what you can see to a single focused slice. A computer adds several single slice images together to reconstruct them into a 3-D image on a monitor. Since the microscope aperture blocks a lot of light, a high intensity light source such as a laser is needed to produce enough light from the specimen.

Several different imaging modes are used during confocal microscopy to study a variety of specimen types. They all rely on the microscope's ability to create high-resolution images, called *optical sections,* in series through fairly thick sections or whole specimens. The resolution of an LSM is about 1.2 nm, and it can be used along with standard fluorescent microscopy.

LSM microscopy has several advantages over common optical microscopy, including a controllable depth of field (resolution comes only from focused light); ability to cut out image degradation and out-of-focus information; and capacity to collect a number of optical sections from thick (often opaque) specimens. LSM provides extremely high-quality specimen images and can be used in a variety of research areas.

More Cool Tools

The ability to measure and manipulate individual nanoscale structures has led to big advances in nanotechnology. Scanning probes, high-resolution electron microscopes, and other sophisticated tools have made it possible for researchers to create new structures, measure new properties, and evaluate new applications. Often, the only things slowing down the show are the chemical composition of a nanoparticle and its electronic and thermal characteristics.

We have seen many different kinds of imaging instruments used as nanotechnology has developed, but which are the true workhorses? Which are the go-to instruments for the best resolution and data? The answer depends on the starting material (for example, proteins need different imaging capabilities than ceramics).

Electron microscopes, long the mainstay of science research on the submicron scale, are hard to beat. Various types of electron microscopes are now able to image individual atoms in nanoparticles and materials with subnanometer resolution. Scientists can get data on elemental composition from electron energy loss and X-ray emission measurements almost to the atomic level. New methods have brought about huge advances in the understanding of magnetic nanostructures.

The STM and AFM have kicked the development of new SPMs into high gear. As described, SPMs measure local properties and shapes by bringing a super small and sharp tip near a solid surface. The distance between tip and material surface lets the SPMs operate in environments impossible for traditional vacuum-based surface analytical techniques.

New scanning probes created after the initial instruments' creation, however, make it possible to go beyond topography and look at many other properties, including the following:

- *Electronic structure at low temperatures* (e.g., scanning tunneling spectroscopy)
- *Optical properties* (e.g., near-field scanning optical microscopes) allows optical access to sub-wavelength scales (50–100 nm)
- *Temperature* (e.g., the scanning thermal microscope) uses a temperature-sensing tip to map temperature fields of electronic/optical electronic nanodevices and measure nanoparticles' thermal properties

- *Dielectric constants* (e.g., scanning charge-storing microscopes) enable a researcher to map out problems in semiconductor devices with nanometer-scale resolution
- *Biological molecule folding and recognition* (e.g., nanomechanics) allows single molecule mechanical measurements to give scientists an idea of what is going on structurally (earlier biological systems have mostly been studied through structural averages)
- *Chemical information* provides access to and allows physical, chemical, and biological observation at nanometer scales

NANOMANIPULATORS

Materials as super small as single atoms and molecules can be moved around and used in a nanosized electrical circuit to act as switches at the atomic level. This compositional and structural control is a huge leap forward in materials development. There have been a lot of advances in nanoscale manipulation, and some of the most important ones work with different types of electron microscopes. They are computer-controlled SPM, optical tweezers, and nanomanipulators.

Computer-controlled SPM permits real-time, hands-on human nanoparticle manipulation. One type of nanomanipulator system has a virtual-environment interface to SPMs—something like a virtual reality game. It gives the researcher virtual surface accessibility at about a million-to-one scale. This direct human-SPM interaction provides not only enhanced measurement ability, but special transducers provide a sense of touch or force-feedback, called a *haptic interface*. They are still pretty crude, but in a few years the technology will be ready to do nanofabrication and/or repair of devices and structures.

Optical tweezers provide another way to grab and move nanometer structures around in 3-D. This ability is particularly important in studying atom and molecular dynamics, since molecular biophysics seeks to understand the behavior of single molecules. Optical tweezers make direct observation of structural parameters possible. Optical tweezers work by focusing a light beam on a particle in a liquid. The force of the light is strong enough to keep the particle in one place; if it moves toward the edge of the beam, the light very gently pushes it back toward the center of the focus. Using optical tweezers, researchers are able to understand how a protein or other polymer moves and responds to applied forces.

Nanomanipulators have been developed to be used in SEMs and TEMs by integrating one or more *piezo-controlled* tips (with several controlled directions of movement) into the sample holder.

> A *piezoelectric motor* or *piezo motor* is a type of electric motor based upon a material's change in shape in an electric field.

A piezo-driven TEM specimen holder has been used to study the mechanical interaction between nanoscale crystals and mechanical loading/bending of carbon nanotubes. This type of specimen holder allows the testing of conductance through individual rows of atoms, making it possible for researchers to find the number of atoms in a sample's nanostructure. Nanomanipulators are being used to test the newest electronic circuit boards and integrated circuits, since their tips are small enough to touch the tiny conducting contact pads.

Fabrication

While nanotechnology is looking toward amazing applications in everything from computer design to cancer cures, a lot of complex, associated equipment (such as SEMs, SPMs, and TEMs) is used to study nanoparticles' properties.

Scientists use these instruments to inspect, produce, and model nanomaterials. Nanotechnologists primarily use three main tool types: inspection tools to see nanoparticles and nanomaterials, manufacturing tools to make nanomaterials, and modeling tools to predict nanoparticle characteristics. These tools play an important part in emerging nanotechnology research and applications. The business and investment aspect of nanotechnology tool use and growth is further discussed in Chapter 12.

Theory, Modeling, and Simulation

In the areas of *theory, modeling,* and *simulation* (TM&S), most nanotechnology progress has been connected with the introduction of more powerful computers. Related advances in software, algorithms, and new theories have also been useful. This merging of different kinds of computational techniques (e.g., quantum chemical and molecular dynamics) allows high-fidelity simulations of nanoscale materials.

Although modeling nanoscale systems is complex due to unknown variables at the nanoscale, it is an important developmental step in the following:

- Reducing design time for new materials
- Developing nanoscale devices from new materials (e.g., carbon nanotubes)
- Increasing reliability and predictability of device operation
- Designing and optimizing new nanotechnologies

TM&S can be used to improve complex or averaged nano measurements made under experimental conditions. TM&S is also great for mimicking living cells' systems as models for future nanotechnology applications. In fact, many current nanotechnology device and system designs have been based on our understanding of natural nanoscale mechanisms and proteins.

But it won't happen overnight, nanotechnology TM&S has a ways to go before it's completely successful. For example, quantum theory along with detailed simulations will need to encompass methods to study nanoscale material properties like electrical, magnetic, chemical, and thermodynamics. For nanoscale materials/devices to be improved from modeling, thousands of design alternatives will have to be considered before costly manufacturing can start.

SCALING IT DOWN

The properties of nanoscale devices in larger environments must also be modeled. This modeling will probably have to cross several different measurement scales (e.g., molecular, nano, micro, and bulk) between initial idea to the manufacture of a working device. You can't just focus on the individual parts. Nanoscale devices have to be modeled in the context of their environment. In electronics, modeling progresses in the following pattern of increasing complexity:

Materials ➔ Devices ➔ Circuits ➔ Systems ➔ Architectures

For electronics to move easily into the nanoscale, new tools connecting across different scales will be essential. Better TM&S allows the researcher to simulate electronic devices by treating them as a whole. When the interactions between different parts and functions of devices are improved, then engineering improvements are possible.

Computing Needs

To continue making different nanoscale advances, some modeling areas need work like advanced mathematics in microscale theory and multi-scale methods. Molecular complexity also needs new methods for optimizing complex structures that will work in predicting nanomaterial self-assembly (e.g. protein folding).

These different methods must also readily come together and be easy to use by lots of people in order to impact overall device design. Results, in the right form and with the right accuracy and uncertainties, also have to work together with other math calculations and theory. Collaborative problem solving, as well as shared databases, make it possible for researchers in scattered geographical areas to work on the same problems and develop a collective set of theories and software.

With more interest in simulation, nanotechnology will have another tool for improvement. Massive parallel computers and advanced software raise simulation capabilities. These supercomputer facilities are basic to large-scale modeling and simulations needed to move nanotechnology forward and solve material problems.

Simulation Factors

Scientists at Oak Ridge National Laboratory in Oakridge, Tennessee, and elsewhere are synthesizing, characterizing, and manipulating nanoscale structures. While offering a wide range of basic and applied science opportunities, their work also poses tough experimental questions. Since their experimental data is gathered at the nanoscale, most measurements can't be interpreted without first creating a mathematical model of the interaction between the measuring tool and the measured material/structure. Table 4-1 lists some of the parameters/factors that nanotechnology simulations integrate.

For example, a theoretical model might be needed to define the interaction/location of neutrons and a nanoparticle's atoms. Or to understand the data from an atomic force microscope moving across a surface, a model might be needed to show how the AFM tip interacts with the surface molecules.

Table 4-1   Depending on the application, many factors influence modeling simulations.

Modeling Parameter/ Factor	Definition
actuator	Converts electrical energy into straight-line mechanical energy
analog	Uses smoothly varying physical quantities to represent reality
biochip	Combines biological diagnostics with fluidics and/or electronics
bioinformatics	Science that expresses genetic information as numerical data
Brownian movement	Random movement of super small particles in solution from thermal motion of solvent molecules
Cartesian system	Shows algebraic concepts visually on a plane graph
diffraction	Ability of a wave to change direction or create visual patterns through interference with itself, other waves, or matter
entropy	Randomness of a system, sometimes expressed as heat per unit volume
emergent property	Appearing in a self-organized system of less complex modules and unpredictable from module properties
nanofluidics	Manipulation of nano amounts of fluid (numerical computation or direct process control)
fluorescence	Re-radiation of photons at different wavelengths over time

Table 4-1   (continued)

Modeling Parameter/ Factor	Definition
fractal	Mathematically identical whatever the magnification
geodesic	Straight, rigid strut connecting the centers of two closely packed imaginary spheres
indirect proof	Logically changing to an absurdity as a disproof of initial grounds
isomers	Molecules containing the same atoms arranged in different ways
laminar effect	Effect of gas flow over a solid surface
linear processing	Computation done in sequential steps
monocoque	Structural design in which an object's skin bears mechanical load
optoelectronics	Cybernetics using electricity and light
orbital	Electron's path around an atomic nucleus
oscillator	Generates regular mechanical vibration
parallel processing	Computation that is performed simultaneously
photocatalysis	Light that triggers an irreversible chemical reaction
plastic deformation	Restricted material flow that increases nanoscale strength
pressure	Physical force per unit area
Quantum dot	Nanoparticle that re-radiates at set wavelengths when irradiated by visible or infrared light
quantum effect	Irreducible packet of energy emitted or absorbed at the atomic or subatomic level
quantum tunneling	Transmission of electrons with observed movement through an intervening solid
rastering	Repeated scans of a surface with a sensor or ion beam
R-factor	Measure of a substance's or system's ability to transmit heat
self-assembly	Ability of substance or system to pull itself into a integral structure
spintronics	Controlling an electron's spin, rather than charge
striction	Mutual attraction of adjacent fixed surfaces in relative motion
superconductor	Material that transmits electricity with little or no resistance
thermoelectric	Converting electricity directly into heat or cooling
topology	Mathematics of their physical forms and their transformations
transistor	Electronic device for modulating input signals
translation	Sideways mechanical displacement
wave front	Complete forward edge of a moving wave

Reproducibility

To get high *reproducibility* (ability to repeat a scientific process and get the same results) and product quality in nanostructured materials, many variables must be considered. In processes from wine-making to biochip fabrication, the affects of temperature, pressure, process time, and concentration are critical.

As part of the United States National Nanotechnology Initiative's (NNI) studies, development, theory, modeling, and simulation are thought to have a huge role in nanoscience/nanotechnology. (NNI is discussed in more depth in later chapters.) In fact, theory/simulation along with excellent research have led to several nanoscience advances, as well as feedback into the simulations.

As part of the work being done at the Oak Ridge National Laboratory, a Nanomaterials Theory Institute was established to boost reproducibility theory, modeling, and simulation. The institute's research includes the analysis of inorganic nanomaterials, electron transport nanostructure self-assembly, and enzymatic catalysts.

Studies in materials modeling, theoretical/computational chemistry, and physics are key study areas in determining atomic quantum mechanics for electron movement or electron spin. Theorists simulate atom and molecule spatial behavior by examining and predicting the forces between and around them. Researchers are also looking into simulation methods to explain nanoparticle neutron scattering such as magnetic, metallic, organic, biological, or some combination.

Compatibility

Nanoscience/nanotechnology offers the chance to bring together complex nanoparticles and materials that were once thought to be incompatible. Some of these materials are described in Part Two of this book. For example, nanoscience offers ways for inorganic surfaces such as gold to be chemically bonded to biological molecules. The resultant hollow spheres can be used in a controlled way for targeted drug delivery.

How is this possible? Size! Again this proves that the manipulation of nanoparticles can produce far more new materials than manipulations of bulk materials. Many more possibilities exist in fundamental biology, chemistry, and physics at the atomic and molecular levels. And a nanotechnology researcher has a much greater selection of sophisticated tools at his or her disposal.

Quantum Needs

Work has been done to scale down semiconductor devices to nano-scale dimensions. Transistors with high-end microprocessors are being scaled down to the nanometer world. As a result, the complex kinetic and quantum properties of semiconductors present tough new challenges. Developing quantum and kinetic simulation tools for nanodevices is affected by thermal conductivity of different materials.

Quiz

1. Which microscope uses a laser (UV) light and scanning mirrors to sweep across a fluorescent sample?

    (a) scanning confocal microscope

    (b) atomic force microscope

    (c) light microscope

    (d) scanning electron microscope

2. Along with product quality, what other key factor must be considered in the major manufacturing of nanomaterials?

    (a) cost of materials

    (b) reproducibility

    (c) technician vacations

    (d) thermal cooling

3. The measure of a lens or optical system to form separate and distinct images of two objects with small angular separation is called

    (a) microscopy

    (b) resolving power

    (c) molecular electronics

    (d) photosynthesis

4. Scanning probe microscopes are instruments used to study the

    (a) elemental composition of a sample

    (b) width of a hair follicle

    (c) surface character of materials from the atomic to the nanoscale

    (d) rotation of the Earth on its axis

5. A TEM can see images 1000 times smaller than a compound microscope and about

    (a) 10 times smaller than a human eye

    (b) 50 times smaller than a human eye

    (c) 50,000 times smaller than a human eye

    (d) 500,000 times smaller than a human eye

6. Which of the following tools are not used by nanotechnologists in their work?

   (a) inspection tools

   (b) manufacturing tools

   (c) modeling

   (d) can opener

7. To see C_{60} molecules and other complex molecular structures, scientists use

   (a) magnifying glasses

   (b) lightning bugs

   (c) electrons, instead of light

   (d) binoculars

8. Which microscope works like a phonograph that plays records with a needle tip?

   (a) ATM

   (b) ADD

   (c) LSM

   (d) STM

9. Nanoscience offers ways for inorganic surfaces such as gold to be chemically bonded to

   (a) mosquitos

   (b) aluminum

   (c) carbon dioxide

   (d) biological molecules

10. When a surface dip, hole, or groove causes voltage to dip as the electrons between the probe tip and sample get farther apart, it is called

   (a) burrowing current

   (b) tunneling current

   (c) magnetism

   (d) underlying current

Part One Test

1. Which microscope works like the old phonographs that played records with a "needle" tip in grooves?

    (a) Simple compound

    (b) LSM

    (c) STM

    (d) TEM

2. A nanometer is equal to

    (a) one billionth of a meter

    (b) one millionth of a meter

    (c) one thousandth of a meter

    (d) one hundredth of a meter

3. The strongest molecules known before C_{60} were

   (a) lead

   (b) gold

   (c) krypton

   (d) diamond

4. Magnification is described as the amount that an image is

   (a) reduced under a microscope

   (b) enlarged under a microscope

   (c) solidified under a microscope

   (d) back lit under a microscope

5. The metric system is based on the

   (a) bushel and peck

   (b) meter and kilogram

   (c) pinch and pound

   (d) yard and mile

6. An electric motor based on a material's change in shape in an electric field is called a

   (a) pizza motor

   (b) slinky motor

   (c) rotary motor

   (d) piezo motor

7. Nanotechnology is a hot topic for

   (a) toddlers

   (b) investors

   (c) swimming suit models

   (d) English professors

8. Eric Drexler has made a public effort to retract which of his ideas about future nanotechnology?

   (a) black blob

   (b) magenta splat

   (c) gray goo

   (d) green goo

9. Nanotechnologists primarily use three main tool types except

    (a) modeling tools

    (b) manufacturing tools

    (c) inspection tools

    (d) gardening tools

10. Nano-sized zinc oxide has been used to improve

    (a) sunscreen

    (b) windbreakers

    (c) durability

    (d) the weather

11. Nanoshell's core size and gold coating thickness affect its

    (a) weight

    (b) bounce

    (c) color

    (d) smell

12. J.J. Thomson discovered negatively charged particles called

    (a) protons

    (b) electrons

    (c) quarks

    (d) neutrons

13. Collecting, combining, and shaping atoms/molecules into specific structures by chemical and enzymatic reactions is known as

    (a) bottom-up manufacturing

    (b) top-down manufacturing

    (c) industrial waste

    (d) archeology 101

14. Which microscope provides information on materials' atomic composition, bonding environment, and electrical conductivity?

    (a) MET

    (b) AEM

    (c) AOL

    (d) AIM

15. In 1960, Dr. Richard Feynman emphasized that moving things atom-by-atom was not impossible, just that

    (a) it was a messy job

    (b) tools were not currently available

    (c) it didn't pay well

    (d) it would take forever

16. What devices provide another way to grab and move nanometer structures around in 3-D?

    (a) optical tweezers

    (b) salad forks

    (c) jewelers' pliers

    (d) boxing gloves

17. The simplest structural unit of an element or compound is a

    (a) crystal

    (b) nanotube

    (c) molecule

    (d) polysaccharide

18. Nanotools include all of the following except

    (a) magnetics

    (b) optics

    (c) hammers

    (d) electricity

19. The chemical formula for hydrogen peroxide is

    (a) H_2O

    (b) H_2O_2

    (c) $C_6H_{12}O_6$

    (d) $Pb(SO)_4$

20. Nano has become all of the following except a

    (a) hot research area

    (b) buzz word for merchandisers

    (c) source of media hype

    (d) threat to the universe

21. A nanoparticle attached to a human cell is about the same size difference as a marble sitting on the

    (a) Eiffel Tower

    (b) roof of a Volkswagen

    (c) nose of a dog

    (d) Golden Gate Bridge

22. Nanoscale engineering of surfaces and layers leads to new properties with many of the following properties, except

    (a) optical

    (b) intelligence

    (c) electronic

    (d) thermal

23. *Nano* is the prefix used to show

    (a) 10^{-9}

    (b) 10^{-4}

    (c) 10^2

    (d) 10^7

24. Engineers want to use nanotechnology to build materials

    (a) at the lowest cost possible

    (b) in time for the holidays

    (c) atom by atom

    (d) through string theory

25. The modern Periodic Table contains approximately how many elements?

    (a) 28

    (b) 57

    (c) 82

    (d) 118

26. A molecule's structural formula describes its

    (a) spatial arrangement/placement of elements

    (b) potential cost

    (c) number of carbon atoms

    (d) reactive capability

27. Microscale theory, complexity theory, and multi-scale methods are all types of

    (a) late-night quiz shows

    (b) English grading systems

    (c) theory, modeling, and simulation

    (d) Earth movement studies

28. Ernest Rutherford received the Nobel Prize for chemistry in 1908 and was knighted in 1914 for what work?

    (a) chocolate cake recipe

    (b) discovery of electron spin

    (c) plate tectonic theory

    (d) modern atomic concept

29. Who thought that nanotechnology would make it possible for all the information in the world to be written in a cube 1/200 of an inch wide?

    (a) Richard Smalley

    (b) Richard Feynman

    (c) Robert Curl

    (d) James Heath

30. The smallest things that the human eye can make out are around

    (a) 100 nm

    (b) 1000 nm

    (c) 10,000 nm

    (d) 100,000 nm

31. In 1989, Don Eigler (IBM Almaden Research Center) amazed the scientific community by

    (a) doing research with very little funding

    (b) being the first person to see microorganisms

    (c) moving 35 xenon atoms to spell IBM

    (d) seeking a second Ph.D. in languages

32. Most computer chips are built by what type of manufacturing?

    (a) lateral

    (b) top-down

    (c) topographical

    (d) bottom-up

33. The number of nano papers increased from 0 in 1990 to nearly

    (a) 10,000 in 2005

    (b) 15,000 in 2005

    (c) 20,000 in 2005

    (d) 30,000 in 2005

34. Transmission electron microscopes

    (a) scan the surface of a sample

    (b) transmit electricity

    (c) see samples with light

    (d) look through a sample like a slide projector

35. Buckyballs are spherical, while graphite sheets are

    (a) flat

    (b) dodecahedrons

    (c) diamond shaped

    (d) oval

36. Nanobots are also known as

    (a) a hot new toy this Christmas

    (b) miniature lunar rovers

    (c) molecular assemblers

    (d) a complex card game

37. To be considered nano, a material must have at least one dimension less than

    (a) 25 nm

    (b) 50 nm

    (c) 100 nm

    (d) 200 nm

38. Molecular complexity requires new methods for optimizing complex structures that will work in predicting nanomaterials'

    (a) color

    (b) self-assembly

    (c) smell

    (d) fabrication dates

39. The Guinness Record for the world's largest nanotube model is

    (a) 880 feet

    (b) 1180 feet

    (c) 1420 feet

    (d) 1820 feet

40. Which microscopes can be combined with X-ray and electron spectrometers?

    (a) stereo microscopes

    (b) confocal microscopes

    (c) dark field microscopes

    (d) analytical electron microscopes

PART TWO

Wet Applications

CHAPTER 5

Biology

The nanoscale world is a lot different from the world we know and love, where everything from planes, trains, and automobiles to shoes, pens, and yesterday's cold pizza are governed by their mass properties. For everything measured in millimeters to kilometers, inches to miles, such properties as friction, malleability, adhesion, and shear strength are up against inertia and gravity. Buildings can be built only so tall without falling or becoming structurally unsound.

At the nanoscale, everything is affected by how small you go. Like floating dust particles, nanoparticles are much less (if at all) affected by gravity than larger objects. Instead of being affected by gravity, the state of atoms and molecules is constantly impacted by nearby objects. Outside forces such as magnetism, air or water currents, heat, cold, electricity, and other factors affect a nanoparticle's direction and reactions. Atom-to-atom contact among molecules or objects are stronger than gravity's pull.

As you learned in Chapter 2, nanoparticles exist at the size of single atoms, which are about 0.1 nm wide. A particle or object is considered to exist at the nanoscale when one of its dimensions is between 1 and 100 nm in length. Most biological particles have at least one dimension that falls into this range. Figure 5-1 lists the relative sizes of different unseen particles to give you an idea of how they compare.

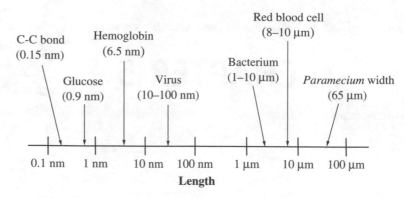

Figure 5-1   Even at the nanoscale, things come in a variety of different sizes.

Most one-celled organisms include organelles, such as a nucleus, mitochondria, and Golgi apparatus, that are even smaller. These structures provide a cell or organism with direction, energy, and the ability to reproduce.

Wet/Dry Interface

The *wet/dry interface* is all about solubility. Much of bionanotechnology offers researchers the ability to change water-insoluble substances into soluble substances that can function and/or react in various aqueous environments, such as a living organism. Finding ways to make the two types (wet and dry) work together is key to using nanotechnology for biological and medical applications.

One way to do this is to take something from the "dry" side, such as gold, and attach it to something from the "wet" side, such as diseased cells, with a specific antibody. In this way, nontoxic particles or even hybrid substances can be used to treat diseases such as cancer. Whether within a cell, organ, microorganism, or ecosystem, nanotechnology has huge potential to affect living organisms.

We can also harness living organisms to create nanostructures. Nature is filled with lots of complex carbon-based machinery designed to carry out chemical, physical, and biological processes. If these biosystems can be harnessed to promote nanoscale synthesis and assembly, the resulting wet nanoscience would change nanochemistry, nanobiology, and materials engineering.

Manufactured nanomaterials are usually foreign materials in biological systems. The way in which they affect biochemical and cellular processes is important in the design of different medical and environmental biosystems. To develop new applications, basic questions about how biological systems (wet) and inorganic nanomaterials (dry) interact must be well understood.

NATURE'S BIOMACHINERY

Nature is versatile in the way it builds everything from the tiniest insects to huge blue whales. Proteins are the foundation of nearly all biomolecular projects. They are composed of a central carbon atom with three attachments: an amino acid group, a carboxyl group, and a side chain of various lengths.

> ***Proteins*** are modular and form long chains of amino acids that fold into specific structures.

Amino acids are added through the *linkages*, in a protein chain. Figure 5-2 illustrates the structure of a typical protein with peptide linkages. The properties of different linkages make it possible for nanotechnologists to work with natural cell mechanisms in biological applications.

In Figure 5-2, you can see that a peptide linkage connecting amino acids has a hydrogen bond donor—the nitrogen-hydrogen single bond group—and a hydrogen bond acceptor—the carbon-oxygen double bond group. The extra carbons in a protein chain carry hydrogen and one of 20 different side chains, usually shown with an *R*. You can see the many different ways that proteins bond affects the potential of various new nanostructures and nanomaterials.

Scientists see nanotechnology as a way to help out Mother Nature in cellular repair. By moving and exchanging atoms and molecules within basic protein structures, scientists can repair and/or eliminate some diseases.

To understand what proteins in cells are actually doing, researchers have to be able to see them. For example, scientists can't unravel the secrets of biomolecules unless basic structures are also understood.

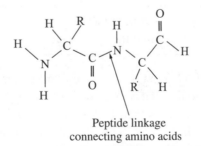

Peptide linkage
connecting amino acids

Figure 5-2   A protein structure can be simple or complex depending on peptide bonds and side chains.

Watson and Crick

DNA (*deoxyribonucleic acid*) proteins make up the *blueprint of life*. They encode an organism's development and growth. In everything from armadillos and aardvarks to newborn baby humans or hippos, DNA lays the foundation of how an organism will look; it is the basis of heredity. It is extremely specific!

In 1951, American biologist James Watson started working with Francis Crick, a physicist at the University of Cambridge in England. Crick was studying the structure of protein molecules with a process called *X-ray crystallography*. Working with nucleotide models constructed from wires, Watson and Crick figured out DNA's structure. In 1962, Watson, Crick, and Maurice Wilkins jointly received the Nobel Prize in medicine/physiology for their work in finding the structure of DNA.

A new understanding of heredity and hereditary disease was possible after DNA's structure was determined. DNA contains patterns for building body proteins, including the different enzymes. Each DNA molecule is made up of two long strands, twisted around each other, connected by hydrogen bonds, and then coiled into a *double helix*. The strands are made up of alternating phosphate and sugar groups held together by hydrogen bonds formed between pairs of organic bases.

The phosphate and sugar groups are called *nucleotides*, which are found in three parts: *deoxyribose* (a five-carbon sugar), a *phosphate* group, and a *nitrogenous base*. Four different nitrogenous bases (guanine, cytosine, adenine, and thymine) are the foundation of *genetic code*. Sometimes written as *G, C, A,* and *T,* these chemicals act as the cell's memory, giving the plan as to how to create enzymes and other proteins. Figure 5-3 illustrates the *alpha-helix* formation of a DNA molecule.

The two DNA chains are held together by *purine* and *pyrimidine* bases formed into pairs. Only specific bases pairs can bond. These pairs are adenine (purine) with thymine (pyrimidine), and guanine (purine) with cytosine (pyrimidine).

Adenine ↔ Thymine
Guanine ↔ Cytosine

Four nucleotides (A, T, C, and G) provide the master blueprint for everything an organism needs in order to live. They copy this information with amazing accuracy. In humans, each cell holds 46 separate DNA molecules, each containing around 160 million nucleotide pairs, yet this massive amount of information is stored and copied with few, if any, mistakes. Better than a lot of computers!

DNA Coding

Human DNA contains the plans, or *code*, for building a human being. Through a very elegant process, a DNA molecule's structure is built by the nucleic acids—A,

Figure 5-3   A DNA molecule is made up of four different nitrogenous bases.

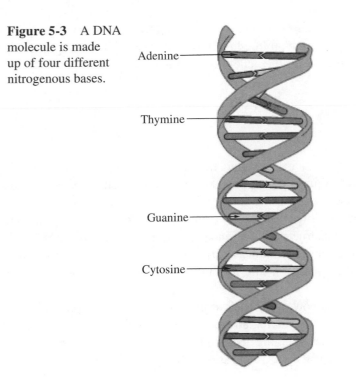

Adenine

Thymine

Guanine

Cytosine

T, G, and C. These then make a copy and bind to other molecules to provide the body's instructions. A typical DNA strand, for example, might be written like this: *AGCGCAAG*. Its complementary strand in the double helix would then be *TCGCGTTC*. This doesn't change unless the protein strand is somehow damaged.

DNA damage can be caused by several factors, such as radiation exposure. This is one reason why X-ray technicians always make sure a woman is not pregnant before undergoing an X-ray. The radiation from the X-ray could cause protein changes within the developing baby's DNA.

Of course, a lot depends on where and to what extent DNA damage occurs during development. If a strand of *GGCAATC* were replicated to *GGGAATC*, it may or may not cause trouble for an individual, depending on what the strand is coding for. One of the great benefits of working with proteins at the nanoscale is the possibility of correcting genetic mistakes/damage. If scientists could move or remove individual, out of position atoms, then medical cures for genetic diseases, such as sickle cell anemia, may be a real possibility. Much work still needs to be done, but new biological pieces of the puzzle are uncovered every day in research.

Some DNA sequences are known to provide control centers for turning processes on and off, some build proteins and other biological materials, and some sequences are only barely understood. Although electron microscopy has been used

for decades, the great benefit of many new nanotechnology tools is that they allow researchers to go deeper (to the atomic level) and see exactly what is going on, as well as how it might be used for the benefit of humankind.

Bioimaging

Imaging in biological systems is basically a matter of sample preparation. The better your laboratory technique and the purer the sample, the better chance you have of discovering structure, composition, and properties. Often, new students in a laboratory will get lower yields (purified sample amounts) using the same purification methods as their mentor/instructor. Experience and attention to detail make a difference! Sometimes something as minor as carrying a sample across the room to a refrigerator compared to putting it into ice right away can lower yield by several percent.

Experienced scientists know every trick to produce the best sample yield possible. From there, it's a matter of taking advantage of a sample's characteristics. If a sample fluoresces, they use imaging techniques that measure fluorescence. If a sample conducts electricity, they test for various charges.

X-RAY CRYSTALLOGRAPHY

X-ray crystallography gives researchers the most detailed information on atomic structure. Crystallographic imaging provides a 3-D map of electron densities and configuration.

First, a pure crystal from a sample molecule is grown. Then it's placed in an intense X-ray beam. The crystal diffracts the X-rays into a pattern to be analyzed by a computer that creates an *electron density map* and shows the location of the crystal's electrons. In a protein like DNA (deoxyribonucleic acid), the location and bonding of amino acids can be imaged. Figure 5-4 shows DNA base bonding.

In fact, atomic positions can be determined within a fraction of an angstrom (10^{-8} cm). Depending on how the crystal is formed, some structural differences may be seen when the protein is in a fluid environment, such as within the cell cytoplasm.

ELECTRON MICROSCOPY

In Chapter 4 the major role of electron microscopy in determining overall structure of biomolecules was described. With average resolution of around 2 nm, these scopes allow scientists to see the overall shape of biomolecular complexes, by not

Figure 5-4 The location of amino acids can be imaged through X-ray crystallography.

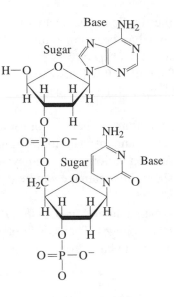

usually single atoms. Scanning electron microscopes (SEMs) and transmission electron microscopes (TEMs) are used together to figure out the complex structures of bionanomolecules.

When electron microscopy details are added to 3-D characteristics seen in X-ray crystallography, computer simulations can be used to analyze and generate the atomic structure of large, complex molecules.

Imaging Single-Walled Carbon Nanotubes

New ways of forming, purifying, and analyzing bionanoparticles are underway at Rice University in the work of professors Bruce Weisman and Rebekkah Drezek at the Center for Biological and Environmental Nanotechnology (CBEN).

> *Spectroscopy* is the analysis of a substance through studying how light of various wavelengths interacts with a substance. The result is called a *spectrum*.

SWNTs would make it easier to get better imaging of biological environments, but a technical challenge must be overcome. When carbon nanotubes are produced in the laboratory, lots of different tubes are produced at the same time. They need to be identified, selectively produced, and separated after initial creation to be useful for biological applications.

At Rice, the first studies of nanotube *spectral* (wavelength absorption) analysis have been done in biological media. This is important to determine which types of tubes work best in specific applications. Data on the energies of 33 distinct semiconducting SWNT types has been collected. Additional progress in telling SWNT types apart will make optical spectroscopy the best method for the purification and analysis of SWNT samples. It will also help scientist create the best SWNT types for bio-imaging, and allow for fingerprint identification of individual tube types in a sample.

One of the most important features of new nanobiology research is the ability for "needle in a haystack" detection and intervention. Beginning when protein damage is first caused to its later effect on various cells and systems, nanotechnology helps track changes. Figure 5-5 shows how damage can be detected at different stages of a process.

CBEN scientists have imaged mouse *macrophage cells* (cleanup structures) nanotubes held within the macrophages. They have noted that the mouse cells seem unbothered by *in vitro* nanotube uptake, even though nanotube fluorescence was

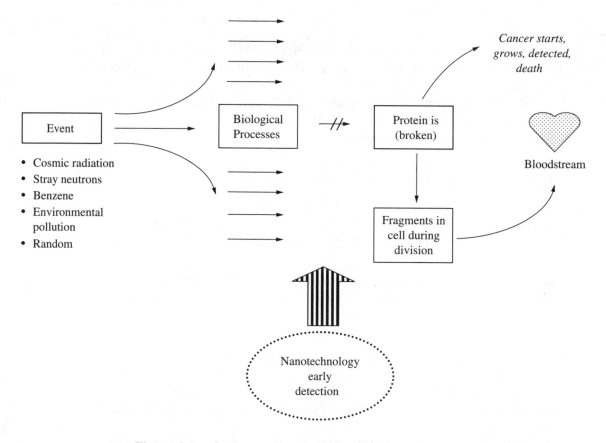

Figure 5-5   Nanotechnology is expected to help with early detection of cell damage and disease.

still visible in the macrophages. Detection of SWNTs will give scientists good ways of tracing nanotube interactions within organisms and may be the starting point for future groups of biological contrast agents and fluorescence markers.

> *In vivo* means within a living organism. *In vitro* means isolated from a living organism (e.g., in a test tube).

Bionanosensors

Regular biosensors have been used for many different applications including health care, environmental monitoring, pharmaceutical discovery, food processing, cosmetics, chemical industries, bioterrorism/defense, and bioprocess monitoring/control.

Bionanosensors are designed to pick up specific biological signals usually by producing a digital electronic signal associated with a specific biological or chemical compound. New methods such as micro/nanofabrication as well as advanced electronics have made development of much improved biomedical sensors possible. These advanced biosensors have the ability to provide big changes in the medical, pharmaceutical, and environmental industries. Individual monitoring nanodevices such as glucose sensors for diabetics are also in development.

Bionanosensors provide scientists with selective identification of toxic chemical compounds at ultra low levels in industrial products, chemical substances, air, soil, and water samples, or in biological systems (e.g., bacteria, viruses, cells, or tissues). By combining very specific biomarkers (e.g., dyes) with optical detection and high-powered computer systems, bionanosensors are able to find and differentiate between complex components.

Most biosensors work by measuring sample interactions with a reactant as it forms in to a product. The reaction is picked up by a sensor that converts it to an electrical signal. The signal is then displayed/recorded on a computer monitor. Reactions within biological processes can be picked up by transducers in several different ways, as listed in Table 5-1. The type of sensor used is often determined by the specific biological process.

The following characteristics are important components of bionanosensors:

- Able to isolate a specific biological factor with little interference
- Quick response time
- Biocompatible
- Super small (nano) in size
- Super sensitive
- Super accurate
- Tough
- Lower cost on more/different tests per sample

Table 5-1 A biosensor transducer can detect changes between reactants and products in many ways.

Transducer Method	Type of Biosensor
Heat output or absorption	calorimetric
Changes in charge distribution	potentiometric
Movement of electrons from a redox reaction	amperometric
Light output or absorption	optical
Mass effects	piezo-electric

Bionanosensor development faces a few hurdles before it can become the norm in most laboratories and hospitals around the world. First, bionanosensors have to be integrated/assessed with regard to current clinical methods, as well as developing bioengineering techniques and advanced electronics. For example, the number of times a biosensor can be used during a process is often limited as proteins build up on the biologically active interfaces.

Next, electronic and biological interfaces between materials need to be joined with various systems to get high sensitivity, selectivity, and stability.

SURFACE TENSION, VISCOSITY, AND CHARGE

At the nano or molecular level, basic properties of fluids (wet things) don't act the same as they do for solids (dry things). This is especially important for bionanomaterials, since the interaction between atoms and molecules is much affected by their bigger (bulk) surroundings. For particular bionanosensors to work well, their interactions with their local environment must be understood.

Surface Tension

Surface tension sounds like something that happens at corporate board meetings and cocktail parties for competing political candidates. Actually, you see surface tension at work every time it rains and water droplets bead up on the leaves of your favorite rosebush or on the newly waxed finish of your car.

> ***Surface tension*** happens when the outside molecules of a liquid pull together to conserve energy.

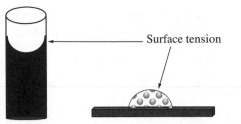

Figure 5-6   Surface tension of water exerts fairly strong forces.

Surface tension (see Figure 5-6) is an attempt by molecules to conserve energy by minimizing the surface area that is exposed to the surroundings. If you think about it, people do the same thing: Whenever we step out of a nice warm building into frosty wintertime air, we pull our hats further shove our hands in our pockets, and pull our jackets tighter around us to conserve energy (body heat) and minimize the skin that is exposed to the frigid wind.

Researchers working to move fluids at the molecular level (actually molecule by molecule) must take surface tension into consideration. *Microfluidics* and *nanofluidics* are new research fields that deal with moving super small amounts of liquids. In order for sensors and other medical applications to work, scientists and technologists must solve the problem of getting liquid samples, such as a drop of blood, to the detectors. Fluids at the nanoscale are affected by surface tension in a huge way and don't easily go where scientists want them to go.

Viscosity

Ever heard the expression "thick as pea soup"? Usually it is used in describing dense, difficult to navigate fog. Anything *viscous* is dense and tough to move through or mix. From the kitchen, honey, molasses, and Uncle Fred's five-alarm chili come to mind as perfect examples of viscosity.

When it comes to food, thick is usually a good thing. But at the nano level, viscosity causes technical difficulties. When researchers are designing sensors and other testing devices, sample viscosity must be considered.

Viscosity is related to another fluid property known as *laminar flow*. For example, when fluid in a container, such as a child's glass of milk, gets knocked over, the spill seems like it involves a lot more fluid that you might expect. Actually, the spill spreads out all over the table, flows over the edges, and drips down onto the chairs and floor. Liquid spreads out a lot because of fluid dynamics and *laminar flow*.

> ***Laminar flow*** is the smooth, continuous flow of the individual molecules of a fluid in a specific direction.

Fluid flow and viscosity are important when designing microchannels for samples to get from the sample site to the testing/measurement sensors. A clogged flow path is not good.

Electrokinetics

The way the engineers get past viscosity and laminar flow problems at the nanoscale is by using electricity. They use their sample's conductivity and electrical characteristics to help move the fluid's molecules in the direction they need to go.

> *Electrokinetics* uses an electrical charge across a super small channel to move molecules along the channel in a specific direction.

Electrokinetics are used in the laboratory in a couple of different ways. One method uses electricity to move or separate molecules in a sample by sending an electrical charge across a channel. This is known as *electrophoresis*. Dyed samples (such as proteins) are placed at the top of a gel substrate and a current is applied. Depending on size and other sample characteristics, molecules move down the gel at different rates. Researchers compare the final locations of the different protein fractions to figure out what the sample is made out of.

Like electrophoresis, electro-osmosis uses an electrical charge to move fluid molecules, but it also charges the sides of the channel itself. The charged particles interact with each other and the wall electrical charges moving down the channel at the same rate.

Biologists have used electrophoresis and electro-osmosis for many years to describe the properties of large (milliliter) samples. Now they have the ability to use much more refined techniques at the molecular level.

Fluid Electrical Force Microscopy

Researcher Dr. Jason Hafner at CBEN has found a way to measure charges at the nanometer scale in the development of biological nanosensors. This new form of microscopy, called *fluid electrical force microscopy* (FEFM), tracks tiny charges on a single molecule. FEFM has been used in a number of biological systems, including the lipids (fats) in biomembranes. Dr. Hafner's lab has used FEFM to observe lipids clumping in the fluid/gel phase regions of membrane-containing compounds.

BIOSENSORS AND BIOMARKERS

Another recent advance in bionanosensors comes from the area of dentistry. Saliva cleans the mouth, fights tooth decay, and, according to scientists at UCLA's School of Dentistry, acts as a window into the body's overall wellness. Scientists have long known that saliva contains a lot of proteins, hormones, antibodies, and other molecular substances. According to UCLA Professor David Wong, the big advantage of diagnostic saliva tests is that they are non-invasive. Saliva is easy to collect and doesn't have the risks, stresses, or invasiveness of blood tests. Needle sticks are not needed.

Dr. Wong and others at UCLA used biosensors to measure elevated levels of four cancer-associated RNA (ribonucleic acid) molecules in saliva and identify between healthy people and those diagnosed with oral cancer with 91% accuracy. With such accurate methods, dental offices may one day be equipped with real-time detectors to diagnose diseases from saliva samples.

According to Dr. Wong, extended research into biomarkers for other diseases such as breast, ovarian, and pancreatic cancers; Alzheimer's Disease; AIDS; diabetes; and osteoporosis may be diagnosed from saliva samples in the future.

The ability to achieve high resolution analysis of biomembranes makes bionanosensors another great tool in the nanotechnologist's toolbox.

BIOCHIPS

In a *DNA sensor*, a single-stranded DNA molecule is used to find a complementary sequence (remember the complementary pairing of A, T, C, G?) among a mixture of other singled-stranded DNA molecules. New biosensors using DNA probes have been developed.

A DNA chip or *biochip* is made by binding many short DNA molecules to a solid surface. This arrangement takes it possible for researchers to analyze thousands for genes at the same time. A biochip is an important tool for discovering how genetic plans are expressed, for measuring biomarker patterns, or for identifying DNA/RNA nucleotide sequences in a biological sample. When nanotechnology is combined with biochip results, repair and/or preventative measures can be started by geneticists or doctors before full-blown problems get a foothold in the body.

The importance of sensor development, as well as different sensor applications, is described in more detail in Chapter 9.

Affecting the Biological World

These are just a few of the latest methods and discoveries being reported from laboratories around the world. The nano world offers ground-breaking answers to questions that science has asked for centuries.

Nanobiology's ability to affect atoms, cells, organs, and the environment for the greater good gives scientists and physicians new ways to thwart illness and ecological disasters. It's like seeing microorganisms under a microscope for the first time and realizing that an entirely new world is waiting to be explored and evaluated. We now have tools like bionanosensors, bio-imaging, biochips, and fluid electrical force microscopy that can help scientists change the future.

Quiz

1. Outside forces such as magnetism, air or water currents, heat, cold, electricity, and other factors affect a nanoparticle's

    (a) cost

    (b) marketing

    (c) direction and reactions

    (d) size

2. Crystallographic imaging provides a

    (a) clarity rating of diamonds

    (b) look at cosmic dark matter

    (c) 2-D map of linear singularities

    (d) 3-D map of electron densities and configuration

3. Spectroscopy is used to study substances by examining their

    (a) spectra

    (b) smell

    (c) saturation point

    (d) size

4. Who was the first person to see live bacteria?

   (a) Alexander Fleming

   (b) Harold Korell

   (c) Antony van Leeuwenhoek

   (d) Richard Smalley

5. Biomarkers for diseases such as breast cancer, Alzheimer's, diabetes, and osteoporosis have been detected in

   (a) ear wax

   (b) saliva

   (c) hair

   (d) tears

6. Modular and long chains of amino acids that fold into specific structures are called

   (a) lipids

   (b) neocytes

   (c) proteins

   (d) protoplasms

7. DNA probe recognition is based upon

   (a) a molecular hybridization process that matches a nucleic acid strand with a complementary sequence

   (b) an elevated temperature

   (c) a phobia about visiting the doctor

   (d) pattern recognition in a random assortment of unbonded lipids

8. Atomic positions can be determined within a fraction of

   (a) a gnat's whisker

   (b) an angstrom

   (c) a micrometer

   (d) a kilometer

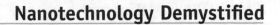

9. Which of the following characteristics are not important components of bionanosensors?

   (a) biocompatibility

   (b) size

   (c) high sensitivity

   (d) high cost

10. Fluid electrical force microscopy involves the

   (a) rising and sinking of bubbles in lava lamps

   (b) mapping of tiny charges across a single molecule

   (c) concept of keeping electrical appliances away from water

   (d) fluorescence of molecule complexes

CHAPTER 6

Medicine

Fifteen minutes after he left, the doctor returns to give you the results of your yearly physical. "Overall, everything looks good," he says, nodding. "However, I was surprised by some unusual activity in your pancreas."

Your smile dims as you wait for the doctor's next words. Your grandfather died of pancreatic cancer, and family stories race through your mind. He had been fairly healthy all his life, except while in his 20s, when he smoked and drank in abundance. You had a glass of wine with dinner last night. Could this be serious?!

Snapping back to the present, you hear the doctor say, "The detected cells and proteins are growing at an accelerated rate. This type of pancreatic cancer...."

Your heart stops. You feel sick. Cancer? How? Why?

The doctor ends a lengthy explanation by saying, "So to nip this early, I'd like you to come in for a couple of immune injections. That should take care of it. You can set up the appointments on your way out. See you next year."

As the doctor leaves, you nod. What a relief! The old organ killer can no longer grow unchecked, thanks to *nanomedicine*.

> ***Nanomedicine*** describes the medical field of targeting disease or repairing damaged tissues such as bone, muscle, or nerve at the molecular level.

By perhaps 2015, this story may no longer be based on fiction; this is our future, as many scientists and physicians envision it. In fact, the United States National Institutes of Health (NIH) Roadmap's Nanomedicine Initiative lists its major goals as finding ways to 1) search out and destroy early cancer cells, 2) remove and replace broken cell parts with nanoscale devices, and 3) develop and implant molecular pumps to deliver medicines. Research to create and use materials and devices at the nanoscale level of molecules and atoms is being heavily funded and pursued.

Nanoscale science can be used to get at medical problems from many different angles, including the following:

- Genetics information storage and retrieval
- Diagnostics, such as the identification of disease
- Detection of overall disease susceptibility, such as Alzheimer's
- Better classification of diseases into different types and subtypes
- Tailor-made drug design based on chromosomal differences
- Gene therapy (e.g., for cystic fibrosis)
- Cell targeting (antibody development that zeroes in on specific cells)

Cancer is still a big killer today. In fact, cancer is not just one disease, but many, each with its own quirks and methods. Cancers use a whole bag of tricks to avoid detection and invade neighboring areas in the body. Figure 6-1 shows the different ways that cancer gets around the body's normal defenses.

Breast cancer, for example, manifests as 14 different types, some nastier and faster growing than others. Although researchers are chipping away at the secrets of different cancers, some are still uncontrollable, especially when found late, after

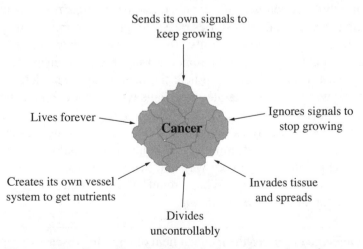

Figure 6-1   Cancer invades the body through a variety of ways.

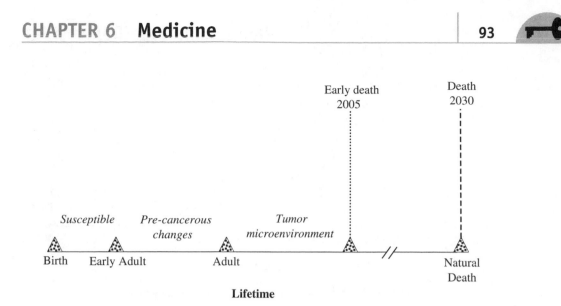

Figure 6-2   Without prevention or medical intervention cancer causes early death.

becoming solidly established in the body. Figure 6-2 illustrates the timeline of an unchecked cancer.

In the long term, nanotechnology and nanoscience may give physicians powerful new tools in the fight against cancer as well as other degenerative diseases—even, perhaps, the effects of aging.

Treatments

Various scientific publications, such as the *Journal of Biomedical Nanotechnology,* focus on nanobiotechnology, biology, medicine, biochemistry, nanoprobes, nanofilters, biomaterials, and biomedical devices, describing the latest in interdisciplinary nanoscale research. The basic nanotechnology techniques being developed by geneticists, molecular biologists, biochemists, bioengineers, and others will open up a diverse and much improved toolbox of treatments for a host of conditions and diseases. Cancer will become a treatable problem, as diabetes is now. Survivability will be enhanced. Figure 6-3 shows how early detection and treatment have a huge affect on survivability.

LAB-ON-A-CHIP

Picture a small plastic chip the size of your thumbnail. Imagine that most of the lab tests that take days to weeks to get results today could be done in a few minutes in your doctors' office using this chip. Is this more far-out science fiction? Not at all. A new nanotechnology concept makes it possible.

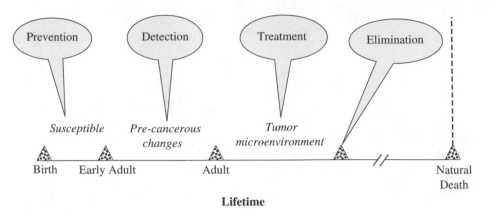

Figure 6-3 A key to treating cancer successfully is timing.

Dr. James Heath, a nanotechnology researcher and chemistry professor at the University of California at Los Angeles (UCLA) is developing a device that will combine 1000 single-cell assays (analysis of a substance used to determine the proportion of isotopes in radioactive materials) on a 1 cm^2 silicon chip. Each chip has rows of cells with individual wells beneath pores in the silicon. Joined to the cell membrane, the pore acts as a channel between the inside of the cell and the outside. Also on the chip are densely packed arrays of *nanowires* (e.g., a line of metal atoms laid down in a row on a silicon base a few nanometers thick). Each nanowire is coated with a bimolecular probe (specific proteins), such as an antibody, which binds to a specific targeted protein. Proteins travel through the membrane and bind to an antibody, changing the nanowire's electrical conductance, which is measured by a detector connected to the array.

Heath and fellow researchers call the method used to create the chip *super-lattice nanowire pattern transfer*. Their technique creates individual semiconductor nanowires that are 8 nm in diameter and 8 nm apart. Earlier attempts by other researchers have produced devices two to three times larger. Figure 6-4 shows details of the first attempts of the lab-on-a-chip idea. Note the much larger sizes used.

Other researchers are examining the amount that intracellular nanoprobes disturb the regular workings of a cell. They want to find out whether results are part of a cell's day-to-day routine or come from changes in the a cell after it has been disturbed. It's a lot like seeing a family on an average day without any extra activities, compared to a day with a soccer game, piano lessons, trip to the grocery store, PTA meeting, and phone call to grandma on her birthday, all before bedtime at 9:00 p.m.

Lab-on-a-chip technology uses methods a lot like those used to make printed computer circuit boards. Chips are made up of micro- or nanofabricated channels through which sample fluids and chemicals flow.

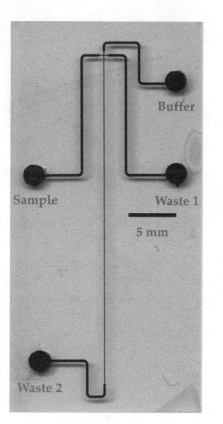

Buffer

Sample　　　　　　Waste 1

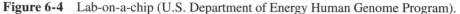

5 mm

Waste 2

Figure 6-4　Lab-on-a-chip (U.S. Department of Energy Human Genome Program).

> *Microfluidics* is the study of using nanoscale liquid-filled channels to move cells to different areas on a base for various different types of study.

Besides being faster, cheaper, and far more accurate than any other available technology, a lab-on-a-chip can be tailor-made to measure different compounds in minutes instead of hours. These tiny chips can also be designed to perform multiple tests simultaneously. Figure 6-5 illustrates how channels of a typical lab-on-a-chip might be arranged to perform a variety of tests using the same chip.

The difference that nanotechnology makes to this system is size reduction and specificity. Although only 16 wells are shown in Figure 6-5, the lab chips of the future could test for hundreds if not thousands of chemical compounds, byproducts, and particles. This would allow for the creation of a universal test for nearly everything from one patient sample. The days of being "stuck" several times for a variety of tests would be gone.

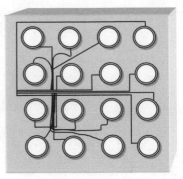

Figure 6-5   Lab-on-a-chip will greatly increase the sensitivity and number of tests possible from a very small amount of sample fluid.

Lab-on-a-chip technology at the nanoscale (e.g. equal to or less than 8 nm apart) would allow biological samples to be mixed, incubated, separated, tested, and processed for data pertaining to hundreds of known conditions, infectious illnesses, and disease states. Cells could be viewed as a set of interconnected activities (e.g., cell signaling, enzyme and nutrient delivery, and the formation of cell products). Instead of just a one-time snapshot of one cell, scientists would have a small window into how everything fits together.

Silicon Nanowires

Harvard University scientists have reported that ultra-thin silicon wires can be used to detect the presence of single viruses electrically. These nanowire detectors can also tell the difference between viruses with great accuracy. If single wires were combined into an array, it is possible that complex arrays could be designed that are able to sense thousands of different viruses.

Dr. Charles M. Lieber, Harvard chemistry professor, has made nanoscale silicon wires that turn on or off in the presence of a single virus. By combining nanowires, a current, and antibody receptors, virus detection is possible.

When an individual virus connects with a receptor, it sparks an electrical change that announces its presence. Researchers have found that the detectors can even distinguish between several different viruses with great selectivity. Using nanowire sensors, physicians will soon be able detect viral infections at very early stages. The immune system would still be able to squash small virus populations, but for particularly dangerous viruses, medical intervention would be available long before the infection would normally be detected by traditional methods.

Targeting Cancer

With medical possibilities beginning to explode, researchers have begun experiments that take advantage of the many benefits nanotechnology has to offer. Nanotechnology is totally multidisciplinary and includes a huge variety of materials and devices from biology, chemistry, physics, and engineering. These techniques use *nanovectors* (hollow or solid structures) for the targeted delivery of drugs to diseased cells, such as cancer cells, and image contrast agents (e.g., that fluoresce or are opaque under a microscope). Some of these are described here.

GOLD NANOSHELLS

Have you ever shined a flashlight through your fingers? As the light passes through the tissues, it looks red, but this is not caused just from blood on the inside, but from light passing through the skin. Long wavelength light can pass right through the skin without too much scattering. This method has been used in photodynamic therapy to treat disease within the body.

Light can be used in different ways. If it hits metal in the body, the metal can get hot enough to cook surrounding tissue (e.g., a tumor). If light hits a particle, causing it to give off highly reactive oxygen molecules, these oxygen molecules will react with the surrounding tissue and destroy it (dooming tumors again).

Researchers Jennifer West and Rebekah Drezek at Rice University's Center for Biological and Environmental Nanotechnology have applied super small particles of gold-coated glass spheres called *nanoshells* created by Rice professor Naomi Halas to improving both the detection and treatment of diseased tissue. Figure 6-6 illustrates the simplicity of gold nanoshells.

> *Nanoshells,* gold-coated silica particles, have tunable optical properties that are affected by size, geometry, and composition.

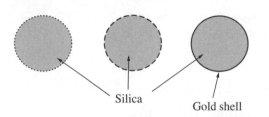

Silica   Gold shell

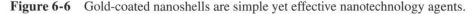

Figure 6-6   Gold-coated nanoshells are simple yet effective nanotechnology agents.

Nanoparticles with a silica (glass) core and gold shell have been designed to absorb light wavelengths in the near-infrared (i.e., in the total spectrum of light) where light's penetration through tissue is greatest. Figure 6-7 shows the different wavelengths of the spectrum. (The visible part of the spectrum is further divided according to color, with red at the long wavelength end and violet at the short wavelength end.)

A new kind of cancer therapy is becoming possible using super small gold nanoshells that travel through a tumor's "leaky" vessels and are deposited. The blood vessels that supply tumors with nutrients have tiny gaps in them that allow the nanoshells to get in and collect close to the tumor. This is called the *enhanced permeability and retention,* or EPR, effect. Nanoshells can also be bonded with targeting antibodies and directed, for example, against *oncoproteins* (cancer proteins) or markers, increasing therapy specificity to the cellular level.

Nanoshells can get to the tumor in two ways: by using a targeting antibody or relying on EPR. Not every cancer has a specific known marker for which an antibody can be designed. Fortunately, EPR means treatment with nanoshells is not limited only to those cancers with specific markers.

To treat breast cancer cells with gold-coated nanoparticles, for example, antibodies are attached to the gold nanoshells, which latch onto the targeted cancer cells. In tests, mouse cancer cells have been treated by shining an infrared laser beam on an affected area. The gold absorbing the infrared light heats up, but the healthy tissues (with no attached gold nanoparticles) keep cool and are not affected. The rising heat (55°C) fries the tumor cells, leaving healthy cells unharmed.

The beauty of this site-specific treatment is that since only the cancerous areas "cook," the rest of the body's healthy tissues are not impacted. This offers a huge benefit over chemotherapy, which kills rapidly growing cells, both friend and foe. (One of the main reasons chemotherapy patients lose their hair is because hair follicle cells divide faster than other cells and are slammed by chemotherapy chemicals.)

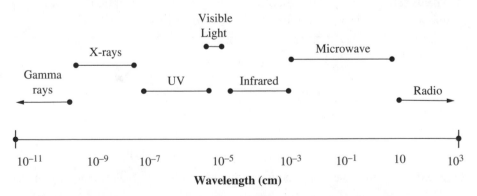

Figure 6-7   Many different wavelengths of light and energy.

In gold nanoshell treatments with mice, scientists achieved a 100 percent effectiveness rate in killing breast cancer cells, compared to the untreated mice, which all died within 30 days. Due to this early success, Rice scientists launched human tests in the summer of 2005.

In a mouse colon carcinoma test, after the intravenous (IV) injection of nanoshells, followed six hours later by illumination of the cancerous area, researchers saw complete destruction of the cancer cells. By day 10, all nanoshell treated tumors had entirely disappeared, while tumors in untreated mice grew unaffected.

After the long-term survival of the mice was tracked, scientists found that all mice in the untreated groups died by day 21, while all the nanoshell-treated mice survived for more than 90 days (all are still happily munching on mouse chow at the time of this writing) with no tumor recurrence.

A researcher investigating a potential new therapy has to make sure the treatment isn't worse than the disease. Problems can arise if these tiny particles don't get where they are needed (biodistribution), if they hang around in the body forever after the therapy (clearance), or if they prove to be poisonous (toxicity). Therefore, nanoshell biodistribution, clearance, and toxicity have also been evaluated, and the nanoshells pass with flying colors.

Such a radically new and improved cancer cure within a body's tissues through infrared light penetration will nearly eliminate a patient's experiencing side effects and suffering. In addition, cure rates will rise considerably since the body's overall defenses won't have to contend with the severe impact of current chemotherapy that has no proven way of telling bad cells from good ones.

TISSUE WELDING

Since nanoshells can be designed to absorb near-infrared light without heating surrounding healthy tissues, their use in laser *tissue welding* in the near-infrared is being examined in the West laboratory at Rice University. Tissue welding allows for better wound healing as well as shorter recovery times. This is especially important in the very young and the elderly.

Nanoshells have been suspended in an albumin (protein) solder and painted onto the cut edges of a wound. When a near-infrared light is shined on the treated wound area, the nanoshells get hot and cause targeted tumor proteins to denature (modify in structure, like frying an egg) and produce tissue welding. Without nanoshells, no welding takes place when the site is illuminated. Figure 6-8 shows how this works.

Although some conventional laser tissue welding (without nanoshells) has been accomplished, it has problems. The wavelengths don't penetrate the tissue very well and welds are thin with a lot of damage to surrounding tissues. The West lab technique minimizes these problems by selecting light wavelengths (near-infrared)

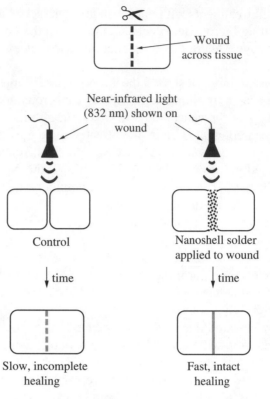

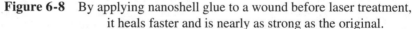

Figure 6-8 By applying nanoshell glue to a wound before laser treatment, it heals faster and is nearly as strong as the original.

that are minimally absorbed by tissues. When nanoshells were suspended in a protein solder "glue" and applied to tissue, then illuminated with near-infrared light, tissue welding happened quickly. Weld strengths were close to the strengths of intact tissue. Early *in vivo* rat tissue welding experiments showed that all welds held wounds closed and got increasingly stronger over the 32-day study.

OPTICAL COHERENCE TOMOGRAPHY

Scattering-based optical imaging technologies, such as *optical coherence tomography* (OCT), are being used for non-invasive cancer diagnostics. As a cancer progresses, tissue refraction changes, making the cancerous tissue look different from healthy tissue when imaged. Early cancer screening, enhanced sensitivity, and potential imaging of molecular markers may also be possible with the use of new contrast agents as mentioned earlier.

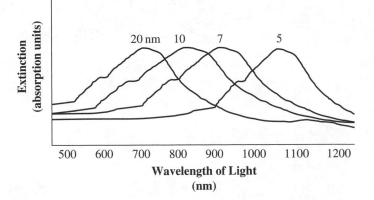

Figure 6-9   Nanoshells can be created to absorb light at different wavelengths.

Since nanoshells have highly tunable optical properties, they can be designed either to absorb *or* scatter light at wavelengths across most of the visible and infrared regions of the electromagnetic spectrum (Figures 6-9 and 6-10). *In vitro* studies have been done using dark field microscopy (which is sensitive only to scattered light). Studies have been done using carcinoma cells with markers (proteins, antibodies, and so on) on their surface. When cells were combined with antibody bonded nanoshells that recognized tumor markers, the dark field cancer cells were easily seen.

These molecular imaging methods can be used with other nanomaterials that have optical properties such as near-infrared fluorescence. Nanoshells can also be designed such that their disappearance is due in part to absorption and in part to scattering.

Bioengineering

The area of engineering that has the highest demand on materials' performance is bioengineering. Nanomaterials are important in medical treatment because their super small size lets them enter into many biological environments and gives them important nanoproperties; plus, their large surface areas act with complex systems to recognize and respond to diseases and tissue damage.

PROTEIN ENGINEERING

Biological systems, including the human body, are made up of lots of proteins. Skin, hair, muscle, blood, organs, eyes, and many other body parts contain thousands of proteins that make up their structure and function. Some diseases, such as sickle-cell anemia and mad cow disease, are caused by damaged protein molecules.

Figure 6-10   Vials of nanoshells of various sizes (Jennifer West).

For many years, scientists have worked to decode the structure of specific proteins. When the human genome was deciphered—a key puzzle piece—the process got a lot easier, and more protein structures became known.

Today, artificial proteins can be created by putting together protein building blocks (amino acids) into the long protein strands that have been identified. Researchers and physicians are able to snip a section from a protein and replace it. The protein can then function as nature intended (or in a specifically crafted way) instead of negatively affecting cell development and function. This is known as *protein engineering*.

> ***Protein engineering*** is the science of making or repairing proteins to benefit medical or agricultural applications.

Nanotechnology can provide even better definition to protein engineering. By seeing exactly how and where a molecular protein structure derails and causes disease, protein engineering methods can become much more important. New scientific fields of *genomics* and *proteomics* are targeting a lot of specific proteins to understand what they do and how their function might be changed or improved to keep people healthy. Someday, artificial proteins that could attack or counteract viral infections may also become a reality.

DNA MOLECULAR THERAPY

Another type of therapy, known as *gene* or *DNA therapy*, makes use of a DNA molecule's ability to *self replicate* (make a copy of itself). It can be used as a kind of biological sensor to find a specific biological particle, membrane, or tissue. Researchers have made DNA strand complements for certain tumor proteins. The *DNA fingerprint* (so-named because the DNA strand is individual and specific) zeroes in on its matching tumor protein in the blood or tissue and binds to it like a lock and key. Since the DNA genes can find only their exact match, the possibility of a mistake is practically impossible.

DRUG DELIVERY

Since the body is a large complex system of different subsystems, it is important that drugs (i.e., medicine) are delivered to spots where they are needed most. Our bodies do this naturally, but scientists are just now beginning to understand how *bioavailability* works and identify the mechanisms that make the body's biological delivery system so perfect.

> **Bioavailability** describes the delivery of healing molecules in the body where they are needed and will do the most good.

Nanotechnology methods of drug delivery use DNA specificity to haul a drug attached to a specific protein to a tumor site to bind to it. Unlike some medicines that affect the entire body, the new drug is delivered exactly to the right spot. This is much safer for the patient since general negative drug interactions are nearly eliminated.

Bioavailability and drug delivery are complex problems—not just a matter of more is better. In the case of toxic chemotherapy drugs, there is often a fine line between killing the cancer cells ravaging the body and killing the patient.

One of the ways the use of nanotechnology has increased bioavailability is the ability to get treatment drugs through cell membranes and into cells. Since most of the replication of viruses and other disease conditions take place within the cell, treatment needs to take place there as well.

Currently, many treatments come to a halt when they get to the cell membrane. They can't pass through because they don't have the correct electrical charge. Putting *polar* (charged) molecules into a *non-polar* (uncharged) membrane doesn't work. One way to get around this is to coat a polar molecule with a non-polar coating that allows it to pass through the membrane and deliver its treatment.

Self-Assembly

In Dr. Michael Wong's laboratory at Rice University, work is underway that allows hollow microcapsules to *self-assemble*. Under certain test conditions, these microcapsules automatically assemble themselves into a hollow sphere, doing the scientist's work for him or her. A microcapsule, shaped like an *o*, can have antibodies or other proteins stuck on the outside, with enzymes or other molecules inside.

Self-assembling microcapsules offer physicians a great tool in drug delivery for a number of conditions and diseases. Using microcapsules to treat a disease such as Alzheimer's, for example, is a four-step process (Figure 6-11): 1) hollow microcapsule spheres with certain size "pores" in their outer coatings are made to self-assemble through chemical reaction with a polymer gel and a salt; 2) larger molecules are encapsulated through chemical reaction at about the same time; 3) smaller reactant molecules are added that slip inside the microcapsules and react with the encapsulated molecules; and 4) micro size medicines flow back out in a time release fashion.

Depending on which molecules are encapsulated, doctors would be able to change the strength and type of medicine prescribed. In this way, treatment dose can be controlled by the amount of small reactant molecules, pH, or temperature.

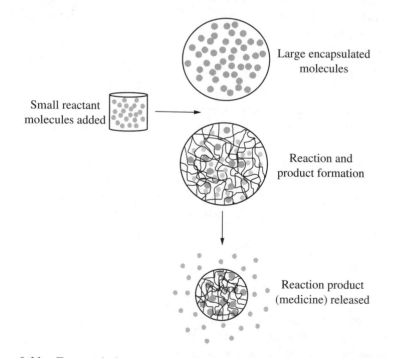

Figure 6-11   Encapsulating reactants allows for time-delivered medicine release.

Multi-Functional Therapeutics

Physicians are always looking for more than one way to approach a problem. So it comes as no surprise that they want more than one way to deliver a mortal blow to diseases such as cancer and sickle cell anemia.

> ***Multi-functional therapeutics*** are medicines that can be delivered to specific areas of the body in different ways (e.g., via mouth or blood).

Nanotechnology can be used to create nanoparticles that deliver their load of medicine in places that regular medicines can't penetrate or have difficulty penetrating. For example, when a patient takes a pill by mouth, the medicine runs into the harsh acidic and digestive environment of the stomach. Next, it encounters the physical barrier of intestinal absorption and length. There's no guarantee that the medication can actually get where its going in concentrations that will help. Figure 6-12 illustrates this oral delivery obstacle course.

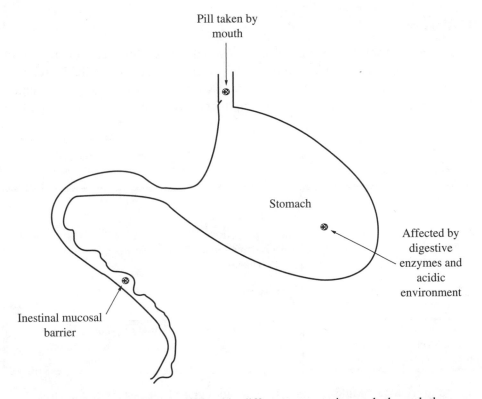

Figure 6-12   Medicine is affected in different ways as it travels through the digestive system.

For years, physicians had few targeting tools other than surgery to use on cancers of the brain. Now, with the development of nanoscale injectable *nanovectors*, they have tools that can cross the blood brain barrier. With the help of MRI and targeting nanovectors, a physician can see during a surgery whether all of a tumor has been removed. Bad cells are targeted by the nanoparticles like lights on a Christmas tree, so that healthy tissue is obvious and a lot less medicine needs to be taken to ensure a cure.

Since diseases can be as individualized as the people who have them, multi-functional therapeutics are important in treatment. For example, they allow the controlled release of medications over days, months, or years. Computer modeling and testing are done to see which method works the best and would be most applicable for a specific site.

Additionally, bacterial/viral infections are known to transform or mutate to get past treatment methods. As they change, it is important that different therapeutics are used to keep infections from getting out of control. Figure 6-13 shows the many advantages an arsenal of multi-functional nanotherapeutics can provide.

IMAGING

Doctors must be able to "see inside" the body to diagnose illnesses properly. Before the most recent nanoscale particle imaging instruments became available, scientists and physicians depended on tissue biopsies, X-rays, magnetic resonance imaging, ultrasounds, and/or blood analyses to help them figure out what was happening in

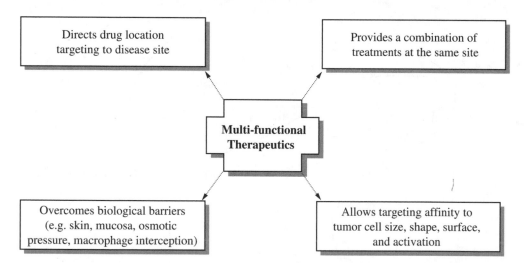

Figure 6-13 Multi-functional therapeutics give physicians more ways to prevent/treat disease.

the body. In the past 15 or 20 years, however, advanced microscopes and imaging equipment (described in Chapter 4) have made diagnosis much less difficult.

Now nanotechnology companies and physicians are teaming up to do *molecular imaging*. Imaging agents are able to illustrate cellular events visually at the time a nanomedicine snags its intended target. No longer is imaging focused only upon large-scale anatomy such as broken bones or hairline fractures.

> ***Molecular imaging*** makes *in vivo* molecular actions visible, quantifiable, and traceable over time in humans or animals.

Imaging agents have been used to provide data on specific regions in the body. Radioactive barium (^{37}Ba), for example, is a contrast agent used to diagnose unusual abdominal pain, gastro-esophageal reflux, cancer, and gastric or duodenal ulcers. Blocked areas appear in white on X-ray images, providing contrast to healthy tissue and allowing radiologists to analyze the digestive tract's inner surface for blockages and tumors.

Nuclear imaging involves injecting a tracer into the body and then tracking its course with a camera. Thallium (^{201}Tl) is a radioactive tracer used to detect heart disease, for example. This isotope binds to heart muscle that is well oxygenated. When a patient with heart trouble is tested, a scintillation counter (detects radiation) measures the levels of radioactive thallium that has bonded to oxygen. Low levels of oxygen in the heart (where very little thallium binding has occurred) is seen as a dark area on a monitor.

The newest nano-imaging tools, often called *probes,* that are being developed are tuned to a specific cellular signal or series of events. This allows physicians to find disease conditions much earlier than ever before. Molecular probes may highlight the first reactions that occur in a disease's progress, long before it shows up in blood work or as a strange lump.

The old saying, "An ounce of prevention is worth a pound of cure" applies here. It is a whole lot easier to fix or stop a problem when it is gnat, than when it's a charging rhinoceros!

NANOSTRUCTURED BONE REPLACEMENTS

When a bone is damaged in a lot of places or across a large area, bone replacement can take a long time to heal. In the case of a person with poor circulation or advanced age, healing can take nearly twice as long as that of a younger person. Speeding up the formation of bone in these instances of catastrophic bone breakage would help get the person back to normal much faster. This is important because the longer someone is laid up from illness, the greater the chance of their muscles and other systems getting weaker. Think of the astronauts in space for many months. They lose a lot of calcium from bones that aren't being used to support their weight and muscles that get very little hard exercise.

Tissue engineering methods may also be used for these types of applications and others. Nanocomposites that mimic collagen have been created from short peptides. These could be added to areas of brittle or healing bone to reinforce its strength and prevent further problems.

> ***Collagen***, the most abundant protein in the body, is formed by many hydrogen and cross-linking bonds between three tightly wound protein strands that combine into long, thin molecules of tough material.

These short peptides self-assemble into collagen's structure and can form polymers end-to-end to build long (hundreds of nanometers) peptide helices. When these collagen-like materials are combined with nanoparticles, nanocomposites for bone tissue engineering will be available for natural bone to use as a support structure in damaged areas.

Osteoporosis sufferers and people with broken bones may also get a lot of support from carbon nanotubes, according to research being done by Dr. Robert Haddon, director of the Center for Nanoscale Science and Engineering at the University of California, Riverside.

Single-walled carbon nanotubes' strength, flexibility, and light weight make them great candidates for "material scaffolding" that can be used to hold up regenerating bone. This type of nanomaterial could lead to improved flexibility and strength of artificial bone, new types of bone grafts, and advanced osteoporosis treatment methods.

With the availability of integrated nanomaterials, healing will be smoother and faster with fewer setbacks, especially in the elderly.

Nanotoxicity

Since nanoscale materials often act in ways that differ from their larger cousins in bulk sizes, their valuable new traits such as super strength or electrical properties raise questions about whether new products incorporating nanoparticles might be unexpectedly risky.

Currently, hundreds of products use nanotechnology. But nanotechnology use is still in its infancy, and nanomaterial production quantities are small. So far, not much is known about the effects of newly invented nanomaterials on humans. The field is pretty wide open and in an information gathering mode. However, nanotechnology toxicologists looking into the risk factors now have a broad roadmap from an Environmental Protection Agency–sponsored panel of experts on how to proceed.

An 85-page-long report, along with supporting documents, was published in October 2005 by *Particle and Fiber Toxicology*, an online scientific journal, and was designed to characterize potential health effects related to nanomaterials exposure. It also provides a screening strategy.

Although the report focuses on toxic impact from nanoparticles in the body, it doesn't talk much about exposure risk, since few instances are known in which people have been directly exposed to nanoparticles. The report highlights the need to characterize the particles in several ways, such as structure, form, surface area, electrical properties, and the possibility of forming clumps that interact with the body in ways that differ from individual particles. It also suggests methods for examining nanomaterials' impact on various internal organs as well as ways to test inhalation, consumption, or contact with nanoparticles.

The report does not provide methods that could explain why nanoparticles might have biological effects. Detailed information on physiological interactions will have to occur collaboratively between science/medical researchers and clinicians.

Medicine of the Future

Historically, Western medicine has developed advanced technologies to treat acute trauma when time was of the essence. Treating wounds quickly and saving lives are the main goals. Western medicine has also been reactive and geared toward finding therapies for diseases often late in their progression.

Compared to Eastern medicine, where prevention is the goal, Western treatment has been more of a knee-jerk reaction to long-term problems. Focusing on prevention, Eastern methods include acupuncture, acupressure, massage, and meditation. They are fairly "low tech" compared to the Western complex, high-cost instrumentation and chemical therapies used to treat trauma and fight advanced diseases.

This chasm between Eastern and Western methods may shift dramatically in the next 10 to 20 years with the nanotechnology techniques currently under development. These technologies will allow people to have their genetic profile assessed as well as have comprehensive molecular testing via blood analysis. Disease and health assessments will be done perhaps as early as birth, allowing physicians to map potential health issues far in advance of their occurrence.

Predictive and preventative medicine will become the norm. Personalized medicine will change the foundations of how medicine is viewed by the public. No longer will economics drive the quality of health care around the world. New nanotechnology-based health care advances will change science research, pharmaceutical companies, education, and society for the better.

Quiz

1. The medical field of targeting disease or repairing damaged tissues such as bone, muscle, or nerve at the molecular level is called

    (a) nanobiotics

    (b) nanocasts

    (c) nanomedicine

    (d) nanoorthodontics

2. Nanoscale-sized, liquid-filled channels that move cells to different areas on a lab chip are called

    (a) microtubes

    (b) microfluidics

    (c) microstraws

    (d) nanostreams

3. When gold nanoshells travel through a tumor's leaky vessels and are deposited, it is called

    (a) metallic plaque

    (b) symbiotic deposition

    (c) optical tomography

    (d) enhanced permeability and retention

4. Gold-coated nanoparticles have been used to treat

    (a) breast cancer cells

    (b) nematodes

    (c) blindness

    (d) foot fungus

5. Nanovectors are

    (a) nanoscale mosquitoes

    (b) surveying tools to measure angles

    (c) used for targeted delivery of drugs and imaging contrast agents

    (d) super small rodent control devices

6. Collagen is composed of many

   (a) vegan approved vegetables and tofu

   (b) hydrogen and cross-linking bonds between three tightly wound
       protein strands

   (c) proteins and lipids that form a circular structure

   (d) symmetrical structures forming a triangular complex

7. Oncoproteins are

   (a) reptile proteins

   (b) good for the diet

   (c) cancer proteins

   (d) kelp proteins

8. In tissue welding, nanoshells are suspended in

   (a) a protein solder "glue"

   (b) gelatin

   (c) lipid compounds

   (d) chicken noodle soup

9. Nanotechnology will change future medical techniques by making
   them more

   (a) time-consuming and expensive

   (b) invasive and undependable

   (c) acute and alarming

   (d) predictive and preventative

10. Dark field cancer cells are easily seen because of

    (a) contrast between light cells and red blood

    (b) increased light scattering from the nanoshells

    (c) their increased size and cell membranes

    (d) better microscope lenses and mirrors

CHAPTER 7

Environment

What comes to mind when you hear the word *environment*? The snow-capped peaks of the Rocky Mountains or the balmy jungles of the Amazon? Spring flowers or colorful changing autumn leaves?

The environment can be great or terrible depending on circumstances. The oceans can offer contented sailing across clear blue waters or 30-foot waves that crush and destroy. Some people have visions of death and destruction caused by environmental factors—people who experienced Hurricanes Katrina and Rita that hit Louisiana, Mississippi, and Alabama in August 2005, for example.

Whether good or bad, most natural environmental events are beyond our control. Seasons change and storms hit. However, one environmental concern, pollution, is often a problem of our own making.

Pollution

Care of the environment is a challenging issue. We can't simply focus on the atmosphere, because the environment is an extremely complex system made up of countless subsystems all working together. Water and soil with their interrelated properties and inhabitants are also part of the equation. Some people think that the environmental cost of progress is unavoidable, while others want to turn back the clock to earlier, cleaner times. The debate will go on for a long time.

You don't have to be an environmental activist to know that polluting the environment is not only wrong, but hazardous to our collective health. Smokestacks belching out all kinds of foul-smelling pollutants will not only give you a headache and take the paint off your car, but they fill the air with cancer-causing chemicals called *carcinogens*.

> *Carcinogens* are chemicals that directly cause or bring about the onset of cancer.

However, knowing whether soil, water, or food is toxic is not an easy feat, since contamination is most often measured in parts per million (ppm) or parts per billion (ppb). Many people believe only what they see with their own eyes, and toxins are visible only with magnification. These days, we not only have to be careful about what we breath and drink, but we have to watch our intake of some fish, such as tuna, because of the mercury buildup in their tissues from contaminated waters. Table 7-1 lists a few toxic chemicals and concentrations found in soil or water.

Table 7-1   An array of different toxic chemicals have been found in soil and water.

Compound	Toxic Levels (ppm)
Arsenic (playground soil)	10.0
Arsenic (mine tailings – toxic)	1320
Diethyl ether	400
Trihalomethane (water)	0.10
Nitrate (water)	10.0
Nitrite (water)	1.0
Silver (water)	0.05
Cadmium (water)	0.005
Mercury (water)	0.002

In most industrialized nations, the air is filled with smoke, particulates, and a variety of toxic chemicals that are caused by human activities and industrial processes. The most common air pollutants are

- Carbon monoxide
- Chlorofluorocarbons (CFC)
- Heavy metals (arsenic, chromium, cadmium, lead, mercury, zinc)
- Hydrocarbons
- Nitrogen oxides
- Organic chemicals (volatile organic compounds, dioxins)
- Sulfur dioxide
- Particulates

Most of us depend on local, state, and national agencies to act as environmental watchdogs, but since pollutants are often colorless, odorless, and hard to detect, their job is not easy.

Acid rain happens when nitrogen oxides and sulfur dioxide settle on the land and interact with dew or frost. Roughly 95 percent of the elevated levels of nitrogen oxides and sulfur dioxides in the atmosphere come from human actions. Only 5 percent comes from natural processes. The main nitrogen oxide sources include the following:

- Burning of oil, coal, and gas
- Volcanic action
- Forest fires
- Decay of soil bacteria
- Lightning

Water pollution is caused by the sudden or continuing, accidental or deliberate discharge of a contaminating material(s). As the planet's population grows, the oceans and marine environments must deal with more human-created pollutants. Water is most often polluted by

- Sewage
- Extracted and burned fossil fuels
- Oil spills
- Fertilizers, herbicides, and pesticides used for crops, lawns, golf courses, and parks
- Deforested and developed land
- Different by-products of manufacturing/shipping

Cultural, political, and economic forces affect the types, amounts, and management of air and water pollutants today. Greater population is just one factor that adversely affects the environment; most causes and effects are quite complex.

Nano to the Rescue

Sometimes it seems that the ills of the environment are too big to handle. Some people give up in the face of these looming giant problems. However, nanoscience researchers see light at the end of the tunnel.

The design and manipulation of atomic and molecular scale materials offers great possibilities for environmental cleanup. Unique properties of new nanoscale materials can produce advances in cleaner energy production, energy efficiency, water treatment, and environmental remediation.

Many projects are also addressing nanoparticle interactions with biological and environmental systems, including how nanoparticles move in and through the environment via nanoscale fluid dynamics. Researchers are trying to determine how different kinds of environmental contaminants bind to or could be transported with nanomaterials through groundwater systems or how cell interactions/toxicity might occur.

Water Purification

Water is critical to human existence. The lack of a clean water supply not only affects health, but contaminated water from animal and human waste and chemical pollution and runoff is especially harmful. Access to clean water is a bigger problem than hunger in underdeveloped, war-torn, or natural disaster areas. In the United States, drinking water standards have been revised and water treatment methods are being changed to meet stricter contaminant standards.

As the planet's population and agriculture needs require greater volumes of potable water, the need for better purification methods have become particularly important. The use of nanomaterials may offer big improvements to existing water purification techniques and materials and may well bring about new ones. They could also supply water treatment and purification in remote areas where electricity is not available.

Engineered nanomaterials are a new class of materials, relatively unknown to most environmental engineers and water treatment workers. However, this is changing. With more and more research on safe, improved, low-cost, and efficient ways to treat water, general water treatment methods will begin to change, too.

CERAMIC MEMBRANES

Membranes and filters of all sizes are used to separate various compounds and chemicals. Depending on their properties, they have greater or lesser success.

In *ultrafiltration*, pressure pushes against one side of an ultrafiltration membrane, forcing water and low molecular weight compounds through its pores. Larger molecules and suspended solids move across the membrane, getting more concentrated as they are blocked because of their larger size.

> A semipermeable, ***ultrafiltration membrane*** system has pore sizes in the range of 0.0025 to 0.01 microns.

Center for Biological and Environmental Nanotechnology (CBEN) researchers at Rice University have developed a reactive membrane from *iron oxide ceramic membranes* (*ferroxanes*). Due to iron's unique chemistry, these reactive membranes provide a platform for removing contaminants and organic waste from water and cleaning them up. Ferroxane materials have even been found to decompose the contaminant benzoic acid.

When using *aluminum oxide ceramic membranes* (*alumoxanes*) as the ceramic nanomembrane material, membrane thickness, pore size scattering, permeability, and surface chemistry can be altered by changing the first layering of alumoxane particles. Figure 7-1 shows the regular design of these membranes. Membrane thermal properties can be changed to create a range of pore sizes.

Nanostructured ceramic membranes treat and purify water both actively and passively. Ceramic membranes could be placed inline within conventional treatment systems for final cleaning of polluted water and air.

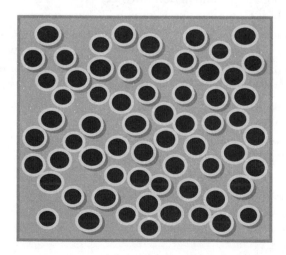

Figure 7-1 Pores in a ceramic membrane.

Figure 7-2 Ceramic nanomembrane. Photo courtesy of M. Wiesner lab, Duke University.

The integration of nanocatalysts into water treatment provides improved options. *Nanocatalysts* are substances/materials with catalytic properties that have at least one nanoscale dimension. Since their greater surface area provides more contact with reactants, they are more efficient than larger materials. These materials could be used in specific applications where contaminated groundwater is already being treated. Nanocatalysts could be activated by common water purification methods, used as treatment additives, and recovered by nanomembranes. Figure 7-2 shows a portion of a nanostructured membrane created from particle templating.

IRON REMEDIATION

Pioneering research by environmental engineer Wei-xian Zhang of Pennsylvania's Lehigh University has shown the potential of iron nanoscale powder that is able to clean up soil and groundwater previously contaminated by industrial pollutants.

Iron, one of the most abundant metals on Earth, might prove to be the missing puzzle piece to the trillion-dollar problem of more than 1000 untreated Superfund (i.e., United States Comprehensive Environmental Response, Compensation and Liability Act [CERCLA]) sites in the United States, other contaminated industrial sites, underground storage tank leakages, landfills, and abandoned mines. The answer seems to come from the fact that iron oxidizes easily and forms rust. However, when metallic iron oxidizes around contaminants such as trichloroethylene, carbon tetrachloride, dioxins, or PCBs, these organic molecules are broken down into simple, far less toxic carbon compounds. Similarly, with toxic heavy metals such as lead, nickel, mercury, or even uranium, oxidizing iron reduces them to an insoluble form that is locked within the soil, rather than being mobile, so they could become part of the food chain and their impacts could be more widespread.

Since iron has no known toxic effect and is plentiful in rocks, soil, water, and nearly everything on the planet, several companies now use a ground iron powder to clean up their industrial wastes before releasing them into the environment. This is great for new wastes, but wastes that have already soaked into the soil and water must be taken care of as well. Enter nanoscale iron particles.

These super small iron environmental dynamos are 10 to 1000 times more reactive than commonly used iron powders. Smaller size also gives nano-iron a much larger surface area, allowing it to be mixed into a slurry and pumped straight into the center of a contaminated site, like a giant IV injection. Upon arrival, the particles flow along with the groundwater, decontaminating the environment as they go. Figure 7-3 shows how the injection process takes places.

Iron particles are not changed by soil acidity, temperature, or nutrient levels. Their size (1–100 nm in diameter and 10–1000 times smaller than most bacteria) allows them to move between soil particles. Laboratory and field tests have shown that nanoscale iron particles treatment drops contaminant levels around the injection well within a day or two and nearly eliminates them within a few weeks,

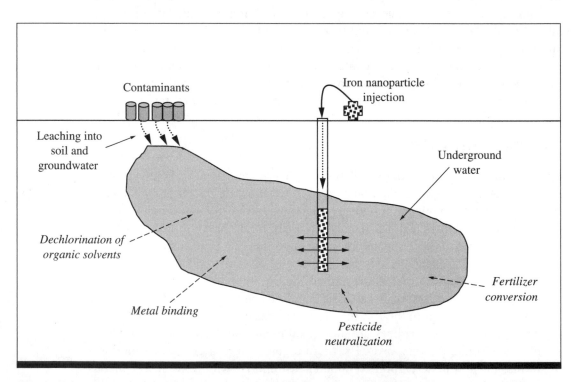

Figure 7-3 Heavy metal contaminants can be locked into the soil with oxidizing iron nanoparticles.

bringing the treated area back into compliance with federal groundwater quality standards. Results have also indicated that the nanoscale iron stays active in the soil for six to eight weeks before the nanoscale particles become dispersed completely in the groundwater and become less concentrated than naturally occurring iron.

This type of nanotechnology innovation can improve environment contamination in a hurry and encourage other researchers to keep looking for new ways to decontaminate other compounds. Zhang's method is also a lot cheaper than digging up contaminated soil and treating it a little at a time, as has been done in the past at polluted Superfund sites.

NANOSCALE POLYMER FLOW

New and improved membrane geometries affect polymer molecules' movement through *nanopores*. The way polymer molecules travel through narrow areas is important to chemical and biochemical processes in both transport and reactivity.

Experiments with genetic protein molecules such as DNA and RNA have shown movement via an electrical current through pores in molecule membranes. DNA and RNA movement is important in the study of viral attachment and in the design of gene-sequencing methods. In fact, because of its structure and shape, DNA is a good model for designing long synthetic polymers. Figure 7-4 shows how a protein can move through a pore in a membrane.

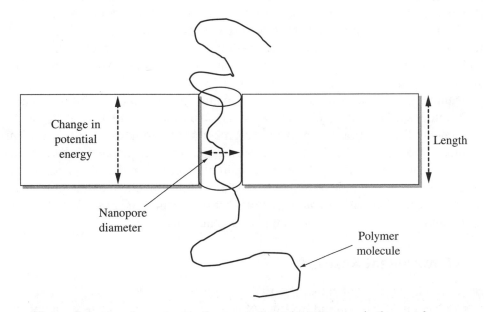

Figure 7-4   A polymer molecule moves through a nanopore in the membrane.

The physics of DNA and RNA movement through nanochannels has a direct effect on future filtration methods and membrane design. Nanoscale computational models also provide data for planning water purification with nanomembranes.

Nanotechnology and Government Research

Nanotechnology is one of the top research priorities of the U.S. government. The nanotechnology research program of the Environmental Protection Agency (EPA) is a part of the government-wide *National Nanotechnology Initiative* (NNI), which provides coordination and direction for this new field. Research and technology taking place at the atomic, molecular, or macromolecular levels (1–100 nm in length) are considered nanotechnology by the NNI.

> According to the NNI, when structures, devices, and systems with unique properties/functions are created, controlled, or manipulated at the atomic level, it is known as **nanotechnology**.

At the nanoscale, the laws of quantum mechanics often create huge changes in the mechanical, optical, chemical, and electronic properties of materials. These properties provide useful and enhanced nanotechnology applications in environmental protection—including sensors for improved monitoring and detection, treatment and remediation methods for low cost and specific site cleanup, green manufacturing that reduces or eliminates waste products, and green energy technology—and the creation of commercially viable clean energy sources. You will read more about energy and nanotechnology in Chapter 11.

Manufactured nanomaterials, however, might also introduce health risks due to their composition, reactivity, and super small size. So it is important that possible nanomaterial interactions with the environment are thoroughly investigated. This includes looking at the effects of natural nanoparticles in the atmosphere, water, and soil, as well as the lifecycles and transport of manufactured nanomaterials. Risk analysis must also include testing of the toxicity of natural and manufactured nanomaterials, exposure paths to humans and other organisms, and possibility for accumulation in the tissues of plants, animals, and humans.

Government Research

Because of toxicity concerns, the EPA has taken the lead in the environmental research applications and implications of nanotechnology. The agency not only has its own research programs, but it also participates in the interagency Nanoscale

Science, Engineering, and Technology subcommittee of the White House Office of Science and Technology Policy, National Science and Technology Council.

The EPA's nanotechnology efforts conducted by the Office of Research and Development (ORD) include the following:

- National Center for Environmental Research has funded research grants worth millions of dollars in nanotechnology applications to protect the environment (e.g., development of low-cost, fast, and simple methods of removing toxic contaminants from surface water; new ultra-sensitive sensors for measuring pollutants; green manufacturing of nanomaterials; and highly selective catalysts).
- Selected research projects are studying the possible harmful effects of manufactured nanomaterials, (i.e., toxicity, transport and transformation, location, and exposure and bioaccumulation).
- Small Business Innovation Research Program awarded contracts to small companies to develop and commercialize nanomaterials and clean technologies (e.g., a SBIR company created an activated carbon nanofiber filter with large surface area that efficiently removes volatile organic compounds and particles smaller than 3 μm from engine exhaust, power generators, and indoor air).
- Several research projects in ORD laboratories are examining nanostructured photocatalysts as green alternatives to hydrocarbon oxygenation; using nanomaterials as adsorbents, membranes, and catalysts to control air pollution and emissions; and testing the effects of ultrafine particulate matter during nanomaterial manufacturing.

NANOCATALYSTS

Although nanofiltration membranes are important in water purification, nanoparticles either in solution or attached to membranes can help ensure that pollutants chemically degrade and don't just travel somewhere else. *Nanocatalysts* are currently being studied for their environmental applications. Catalytic treatments can lower polluted water treatment costs by making it possible for purification methods to be specifically designed to treat chemicals at a particular site.

> *Nanocatalysts* are nanoscale materials that have at least one nanoscale dimension or have been modified structurally to enhance their catalytic activity.

For example, removing various pesticides from groundwater is important. Each case requires a different catalyst and strategy for cleanup. Tailor-made nanomaterial treatment can make cleanup faster and more efficient.

Dr. Daniel R. Strongin, chemistry professor at Temple University in Philadelphia, has used protein structures to design and assemble metal oxide nanoparticles that could be used in environmental remediation. By using nanoparticles made from biological components as nanocatalysts, Strongin and others have been looking at how nanoparticles may be used in environmental remediation (cleaning up polluted areas).

Strongin and his colleagues are studying reactions that would make polluting metals clump or separate out of solution so they aren't transported downstream or soak into groundwater. By experimenting with the reduction of toxic Chromium-6 (on the EPA's groundwater toxic metal list), they were able to combine nanoparticles with Chromium-6 and change its chemistry, which is not water soluble. In the changed form, it is much easier to filter. Strongin's method could help alter toxic metals in lakes, rivers, or streams, and in groundwater for better cleanup.

The group has also been working with other toxic metals, such as Technetium-7, a pollutant at a nuclear waste site in Washington state. Large barrels of aging nuclear waste (from the 1940s and 1950s) are leaking, and groundwater contamination is a big concern. Using the nanocatalysts' unique properties, treatment of Technetium-7 with nanoparticles may accomplish what hasn't been done with larger materials.

ABSORPTION OF CONTAMINANTS

Although common water treatment removes a lot of waste from water, targeted treatment applications are most appropriate for highly toxic compounds that must be removed with high efficiency. For many different types of heavy metals, nanocatalysts won't work, so water treatment must be done with absorption onto polymers or nanoparticles.

For example, arsenic is a common natural and manmade contaminant in water. A poisonous chemical element, it combines with oxygen, chlorine, sulfur, carbon, hydrogen, lead, gold, and iron, and it is present in many rock-forming minerals. It also occurs as a result of geological processes, manufacturing, smelting, and agriculture.

Arsenic poisoning kills by completely disrupting the organism's digestive system, leading to death from poor oxygen delivery to the cells. Poisoning is characterized by reduced cardiac output, rapid heartbeat, lesions, and paleness. Symptoms include violent stomach pains, vomiting, and delirium. It has also been linked to increased bladder and rectal cancers.

Because of its health threat, the EPA has lowered the arsenic standard for drinking water to 10 ppb, a number that can be tough to reach in most water treatment plants.

Globally, arsenic poisoning is a huge problem. In the nations of Bangladesh, India, Mexico, Chile, Argentina, Taiwan, and Thailand, an estimated 10 to 40 percent of the population is afflicted with arsenic poisoning. Although arsenic contamination is generally associated with third-world countries, many drinking water supplies in America exceed the recommended 10 ppb level. New technologies that can target and remove heavy metals such as arsenic from drinking water are critical.

Environmental Exposure Routes

Two elements are involved in risk: *exposure* and *hazard*. Exposure potential predicts the chance that a given system will come in contact with a pollutant in high enough concentration to cause a problem. Hazard describes the problem (For example, exposure to single-walled carbon nanotubes in aerosol form after being inhaled. A follow-up experiment might measure any potential damage to lung tissue after inhalation of single-walled carbon nanotubes.) Chapter 13 describes risk assessment in greater detail.

Overall, almost nothing is known about the toxicology of engineered nanoparticles produced through "wet" interactions. While a lot of information has been gathered on the effects of aerosols on living systems through inhalation exposure, toxicological effects of engineered nanoparticles on biological reactions is still a mystery.

Nanoscale quartz, titanium, and iron oxides are being studied in both cell culture and lung studies to determine how particle size and surface chemistry influence nanoparticles' biodistribution and biological effects. Testing at the National Institute of Environmental Health Sciences and the Food and Drug Administration is being conducted using similar materials, but studies are looking at skin exposure routes for engineered nanoparticles and immunotoxicology.

One of the biggest questions in the minds of many scientists, policymakers, and industrial researchers concerns the consequences of incremental or large release of various nanoparticles into the environment. That is not to say that such a release would create a huge problem—it is just that not enough is known about nanomaterials' interactive properties to predict whether they will have a negative impact.

Much is being discovered about all the positive things that nanomaterials can do. However, having good precautions in place to protect people and the environment is also essential. We must be responsible when using any new technology.

The actual dose of nanoparticles that a biological organism (bacteria, fish, human, and so on) might come up against in the environment is being studied intently. To predict the speed and efficiency of nanoparticles' movement through water and soil, scientists must try to get a handle on nanoparticles' transport methods. Nanomaterials are transported very differently in different porous materials. Among carbon-base nanomaterials, nano-C_{60} and surfactant-dispersed SWNTs show the greatest mobility. In fact, it appears that as nanoparticles build up, particle transport is slowed a bit, like a log jam. Professor Mark Wiesner of Duke University and his colleagues have been studying various nanomaterials for particle transport and toxicity.

GREEN NANOTECHNOLOGY

Nanotechnology has the potential to change manufacturing processes in two ways. The first of these involves using nanotechnology for efficient, controlled manufacturing to quickly reduce waste products. The second makes use of nanomaterials as catalysts that can raise manufacturing efficiency by dropping or getting rid of toxic materials and/or dirty end products.

> *Green nanotechnology* includes technologies that make use of principles of environmentally safe chemistry and engineering.

Green nanotechnology ideally will improve industrial processes and material requirements, develop new chemical and industrial procedures, and replace current unsafe compounds and processes, making advances in energy production and various materials possible.

Nanotechnology research can lead to these technological and environmental marvels in several different ways:

- Atomic-level synthesis of new and improved catalysts for industrial processes
- Inserting information into molecules (such as DNA) that build new molecules
- Self-assembling molecules as the foundation for new chemicals and materials
- Building molecules in microscale/nanoscale reactors
- Increasing alternative energy efficiency through advanced solar cells, fuel cells, energy transmission
- Improving manufacturing to require less energy than current processes

To emphasize the importance of *green chemistry* and manufacturing, the 2005 Nobel Prize in chemistry was awarded to Robert H. Grubbs of California Institute of Technology, Richard R. Schrock of MIT, and Yves Chauvin of the Institut Français du Pétrole in Rueil-Malmaison, France, for their pioneering work on *metathesis*.

Metathesis takes place when the order of a reaction is changed—i.e., when cations ($+$) and anions ($-$) exchange partners, it is called metathesis. Metathesis reactions include precipitation, neutralization, and gas formation reactions.

Here is an example of a metathesis reaction:

$$NaCl(aq) + AgNO_3(aq) = AgCl(s) + NaNO_3(aq)$$

This reaction, a central part of the chemical industry in the development of pharmaceuticals and of advanced plastic materials, has greatly advanced the cause of green chemistry. Metathesis not only uses a lot less energy than earlier methods, but it also decreases the amount of hazardous waste products.

International Council on Nanotechnology

Currently, an international effort to ensure responsible development of nanotechnology has been established to address and decrease risks from nanomaterials. This organization, initiated by the CBEN, is called the *International Council on Nanotechnology* (ICON). It seeks to assess, communicate, and reduce environmental and health risks associated with nanotechnology while at the same time offering benefits to society.

ICON has launched the world's first online database of scientific findings related to the benefits and risks of nanotechnology. The environmental health and risk analysis database is the initial effort to pull together the huge amount of diverse scientific information on the impacts of nanoparticles. The database was developed by the combined work of Rice University scientists, the United States Department of Energy, and the chemical industry.

ICON includes not only industries, but also government agencies, non-governmental organizations, colleges, and universities. This broad partnership will study and assess not only the safety and risk of nano-cell interactions, development of nanomaterial standards, and terminology, but also risk perception and communication. Broad support for this organization's intent will allow it to serve as a central depository for world-wide health and environmental information on nanomaterials.

LOOKING BEFORE WE LEAP

It is possible that these novel materials could pose risks to workers, consumers, the public and the environment. Responsible researchers believe that both the public and private sectors must thoroughly study and address the potential risks of this important new technology to ensure its responsible development.

Quiz

1. Contamination is most often measured in

   (a) parts per hundred

   (b) parts per billion

   (c) parts per quark

   (d) parts per trillion

2. Arsenic poisoning disrupts which main body system

   (a) digestive

   (b) reproductive

   (c) neurological

   (d) auditory

3. Roughly what percentage of elevated nitrogen oxide and sulfur dioxide levels in the atmosphere come from human actions?

   (a) 52%

   (b) 75%

   (c) 95%

   (d) 100%

4. Which technologies make use of environmentally safe chemistry and engineering?

   (a) orange nanotechnology

   (b) micronanotechnology

   (c) green nanotechnology

   (d) tree nanotechnology

5. Which of the following does not describe Technetium-7?

   (a) it is a contaminant at a nuclear waste site in Washington state

   (b) it is used as an additive in windshield wiper fluid

   (c) it is a toxic metal

   (d) it is a source of groundwater contamination

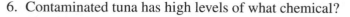

6. Contaminated tuna has high levels of what chemical?

   (a) arsenic

   (b) molybdenum

   (c) strontium

   (d) mercury

7. Toxicity testing of natural and manufactured nanomaterials, exposure paths, and bioaccumulation potential are all important in

   (a) risk analysis

   (b) shoe manufacturing

   (c) cooking classes

   (d) high school locker rooms

8. Materials with at least one nanoscale dimension or that have been structurally changed to enhance their catalytic activity are called

   (a) bioceramic membranes

   (b) space dots

   (c) nanosparklers

   (d) nanocatalysts

9. Chemicals that directly cause or bring about the onset of cancer are known as

   (a) antigens

   (b) leukocytes

   (c) carcinogens

   (d) enzymes

10. ICON is an acronym for

   (a) Indentured Committee on Neighbors

   (b) Indivisible Council on Nanobots

   (c) Incredible Calmness of Night

   (d) International Council on Nanotechnology

Part Two Test

1. Friction, malleability, adhesion, and shear strength of bulk compounds are affected by

   (a) humidity and temperature

   (b) surf and sun

   (c) inertia and gravity

   (d) hydrostatic pressure and elevation

2. Crystallographic imaging provides a 3-D map of

   (a) atmospheric auras

   (b) electron densities and configurations

   (c) mountainous regions

   (d) tree growth rings

3. Nanoscale science can be used for all of the following medical problems except

   (a) genetics information storage and retrieval

   (b) cell targeting

   (c) tailor-made drug design based on chromosomal differences

   (d) ingrown toe nails

4. The most abundant protein in the body is

   (a) nitrogenase

   (b) keratin

   (c) amylase

   (d) collagen

5. "Wet" molecules are found in

   (a) swimming pools

   (b) living organisms

   (c) rocks

   (d) radioactive waste

6. Artificial proteins may be created through protein engineering that could

   (a) make bridges stronger

   (b) increase research funding

   (c) make weather forecasting more accurate

   (d) attack or counteract viral infections

7. Chemicals that directly cause or bring about the onset of cancer are called

   (a) irritants

   (b) bases

   (c) allergens

   (d) carcinogens

8. *Micrographia*, a book on the characteristics of microorganisms, was written by

   (a) George Bennett

   (b) Robert Hooke

   (c) Norman Beckman

   (d) Jack Showers

9. Super small gold-coated glass spheres, used to improve the detection and treatment of diseased tissue, are called

   (a) nanoshells

   (b) oyster shells

   (c) shell tubes

   (d) bioshells

10. What percentage of elevated atmospheric nitrogen oxides and sulfur dioxides come from human actions?

    (a) 15%

    (b) 42%

    (c) 70%

    (d) 95%

11. Pore size scattering, permeability, and surface chemistry can be altered by changing the initial particle layering in what kind of nanomembrane?

    (a) alumoxane

    (b) silicon

    (c) semi-permeable

    (d) biochemical

12. Bionanosensors are designed to

    (a) indicate the presence of senior scientists

    (b) detect important biological signals

    (c) measure rainfall amounts

    (d) determine shoe size

13. Toxic heavy metals such as lead, nickel, mercury, or even uranium in the soil can be neutralized by

    (a) reducing zinc

    (b) adding sodium

    (c) oxidizing iron

    (d) adding chlorine

14. Water pollution is caused by the sudden/ongoing, accidental/intended, discharge of

    (a) sand

    (b) ice cubes

    (c) water lilies

    (d) contaminant materials

15. How many different types of breast cancer have been identified?

    (a) 11

    (b) 14

    (c) 15

    (d) 17

16. Long, modular chains of amino acids that fold into specific structures are called

    (a) lipids

    (b) rhizomes

    (c) origami

    (d) proteins

17. When structures, devices and systems with unique properties/functions are controlled or manipulated at the atomic level, it is known as

    (a) aerodynamics

    (b) microbiology

    (c) nanotechnology

    (d) bioengineering

18. The medical field of targeting disease or repairing damaged tissues such as bone, muscle, or nerve at the molecular level is called

    (a) nanomedicine

    (b) obstetrics

    (c) nuclear medicine

    (d) rhinology

19. The first green algae to be seen was

    (a) *Leptospira*

    (b) *Saccharomyces*

    (c) *Spirogyra*

    (d) *Aspergillus*

20. The delivery of drug molecules within the body and where they will do the most good is called

    (a) chemosynthesis

    (b) pharmacology

    (c) photosynthesis

    (d) bioavailability

21. The first tiny spaces inside a piece of cork were called

    (a) nits

    (b) cells

    (c) chips

    (d) holes

22. With great accuracy nanowire detectors can tell the difference between

    (a) metals

    (b) inorganic molecules

    (c) viruses

    (d) earthworms

23. Because of greater surface area, nano-iron particles are how much more reactive environmentally than commonly used iron powders?

    (a) 2–3 times

    (b) 5–7 times

    (c) 10–1000 times

    (d) not more reactive than regular iron

24. Historically, Eastern medicine has been preventative in nature, while Western medicine was better at treating

    (a) wrinkles

    (b) acute trauma

    (c) stress

    (d) swimmer's ear

25. Magnetism, air or water currents, heat, cold, electricity, and other factors affect a nanoparticle's

    (a) name

    (b) marketability

    (c) investment potential

    (d) direction and reactions

26. The science of making or repairing proteins to benefit medical or agricultural applications is called

    (a) lipid engineering

    (b) home healthcare

    (c) protein engineering

    (d) having a green thumb

27. Within the United States government, NNI is an acronym for the

    (a) Neonatal Nanny Initiative

    (b) National Neurological Institute

    (c) Nanoscience and Nanotechnology Institute

    (d) National Nanotechnology Initiative

28. Reactive ceramic nanomembranes that remove contaminants/organic waste from water use which element?

    (a) potassium

    (b) iron

    (c) lead

    (d) zinc

29. The analysis of samples by studying spectra with an optical instrument is called

    (a) spectroscopy

    (b) climatology

    (c) astronomy

    (d) virology

30. When microcapsules automatically round up into a hollow sphere, it is known as

    (a) a donut

    (b) cryogenesis

    (c) self-assembly

    (d) calcification

31. At the nanoscale, huge changes in mechanical, optical, chemical, and electronic material properties are affected by the laws of

    (a) local and state governments

    (b) photodynamics

    (c) supply and demand

    (d) quantum mechanics

32. When scientists perform tests within a living organism, it is known as

    (a) *in centro*

    (b) *in vivo*

    (c) *in livo*

    (d) *in vitro*

33. Materials that have at least one nanoscale dimension or have been modified structurally in order to enhance their catalytic activity are called

    (a) biomarkers

    (b) targeting enzymes

    (c) nanocatalysts

    (d) antibodies

34. Reactions within biological processes can be picked up by

    (a) transducers

    (b) rubber gloves

    (c) feather dusters

    (d) delivery people

35. Nanotechnology can give physicians powerful new tools in the fight against all but which of the following?

   (a) cancer

   (b) thumb sucking

   (c) degenerative disease

   (d) aging

36. Green nanotechnology includes technologies that make use of

   (a) environmentally safe chemistry and engineering

   (b) natural dyes and fibers

   (c) tectonic processes

   (d) heavy metals such as arsenic, chromium, and cadmium

37. Lab-on-a-chip is accomplished by

   (a) using very small tubes and centrifuges

   (b) super-lattice nanowire pattern transfer

   (c) building labs with very precise tools

   (d) super-small pipettes and wire loops

38. Tests show that nano-iron particle treatments drop contaminant levels around an injection well within

   (a) 3–4 hours

   (b) 1–2 days

   (c) 5–6 days

   (d) 3–4 months

39. When foreign particles are actively ingested and held within a cell's macrophages, it is known as

   (a) lithification

   (b) desalination

   (c) phagocytosis

   (d) halitosis

40. The law that states that the amount a spring stretches is proportional to the amount of weight hanging from it is know as

   (a) Heifer's Law

   (b) Hood's Law

   (c) Hermione's Law

   (d) Hooke's Law

PART THREE

Dry Applications

CHAPTER 8

Materials

Nanomaterials are great and strange at the same time. Carbon nanotubes, for example, add strength, flexibility, and heat protection to plastics, ceramics, and metals. Nanomaterials don't break easily when dropped or smashed. When sliced, nanomaterials "heal" themselves by linking back together. Nanomaterials offer engineers a brand new bag of tricks to make life better for everyone.

Nanomaterials have amazing and useful properties with many structural and nonstructural applications. But they are not new—nanomaterials have been important in the materials field for a long time; we just couldn't see or manipulate them. Gold nanoparticles were used in medieval stained glass, and nanoparticles of carbon black have been used to reinforce tires for nearly 100 years.

Alchemy

Early chemistry, or *alchemy* as it was known hundreds of years ago, was a mixture of trickery and art. It promised amazing things to those who understood its power. However, much of it was a lot of smoke and mirrors.

Some alchemists, known as *adepts*, claimed that through spiritual transformation and feeling the Earth's vibrations, they could achieve personal perfection and the power to create gold. *Puffers* (alchemists who used bellows to fan their forge fires) sought riches through the *transmutation* of metal and used flashy, seemingly magical methods. Puffers used different kinds of furnaces and bellows along with special fuels of oil, wax, pitch, peat, and animal dung to create "gold." It was thought that the hotter the fire, the faster the reaction occurred. It made a good show for the locals anyway.

Although alchemists used odd methods, they did come up with some ideas that helped advance early chemistry. For example, they correctly realized that color was a basic property of an element. In their attempts to make gold, they worked to get a yellowish or golden color on a metal's surface. Using a lime, sulfur, and vinegar solution, which was mixed with whitened copper and heated, alchemists created a golden color that they insisted was newly formed gold.

During this time, crystallization and distillation of solutions were gradually understood and standardized. Previously unknown elements and compounds were also discovered.

Today, researchers are quickly uncovering the unique physics, chemistry, and biology of nanoscale materials and reactions. Researchers are learning about it through physical principles, theoretical explanations, and experimental methods.

Nanotechnology applications come from testing new materials, properties, and processes. Research is uncovering new ways to improve existing commercial products between nanoscale atoms/molecules (~1 nm) and bulk materials (>100 nm). Compared to larger scale methods, the nanoscale is not just another rung on the miniaturization ladder; it is a *qualitatively* new scale. Nanosystems exist within the quantum realm of matter and are becoming measurable through advanced microscopy and other tools. So instead of being science fiction, quantum properties are the actual enabling tools of nanotechnology research.

Smart Materials

Nearly every industry, including biomedicine, energy, chemicals, and electronics, is affected by the nanoscale. Nanoscale materials/processes are responsible for the behavior of materials. Many applications have just been waiting for the right materials to come along—such as economical solar cells or super efficient electrical lines. These materials may make global energy problems a chapter in old history books. As other times of progress (Iron Age, Bronze Age, Industrial Age, and Information Age), with nanomaterials, we are entering a new age—the *Molecular Age*.

Nanomaterials such as carbon nanotubes or nanoshells have superpowers compared to regular carbon or silica particles. Carbon nanotubes, for example, have 100 times the strength of steel, conduct heat better than a diamond, and carry electricity better than copper. Similarly, nanomaterials such as buckyballs, single-walled nanotubes (SWNTs), nanoshells, quantum dots, and microcapsules have been called *smart materials*, and their versatility has not been lost on the science and engineering communities.

Like science, engineering focuses on several research areas, such as aerospace, biomedical, chemical, electrical, environmental, mechanical, and nuclear. However, all areas share a common denominator: advanced materials. Engineers work toward improving or understanding the specific properties of materials. Since nanomolecules and nanotubes were discovered, scientists and engineers have rushed to test all the possibilities these materials offer. Everyone wants to find new ways to use them.

For example, plastics research made possible everything from storage containers and toys to contact lenses and artificial joints. Plastic was the new big thing in the 1950s and 1960s. It changed the way people lived. Lots of things became so cheap to make that they were just thrown away when they got dirty or scratched. (The benefit of this disposable mindset is still being debated, however.)

Nanomaterials seem to be heading in the same direction, but perhaps in a more environmentally tolerable way. Their special properties at the molecular level make plastics seem prehistoric by comparison. Table 8-1 compares the various properties of different nanomaterials.

Engineers now have even greater strength, heat conductance, molecular changeability, electricity transmission, and flexibility to work with. Not only will new solutions to old problems such as disease be found because of their research, but completely new products (unimagined today) are possible. As these products become available, we'll wonder what took us so long to think of them. When technology allows us to mix and match atoms, anything is possible!

Table 8-1   Nanomaterials offer different advantages depending on the application.

Nano Property	Organic Polymers	Metals	Semi-conductors	Ceramics	Carbon
Optical		☆☆☆	☆☆☆		☆
Mechanical	☆	☆		☆	☆☆☆
Electrical		☆☆			☆☆☆
Magnetic		☆		☆☆☆	
Catalytic				☆☆	☆
Absorptive	☆☆☆				

CARBON NANOTUBES

Single-walled carbon nanotubes (SWNTs) are incredibly promising nanomaterials. Their remarkable material properties, such as strength, rigidity, durability, chemical vigor, thermal conductivity, and (perhaps most importantly) electrical conductivity, make them very versatile. Depending on their exact molecular structure, some nanotubes are semiconducting, while others display true metallic conductivity. This ability, combined with their nanoscale geometry, makes them great candidates for wires, interconnects, and molecular electronics devices.

For years, great strides in SWNT development were slow, because of processing problems. This may no longer be an issue since Professors Richard Smalley and Matteo Pasquali at Rice University's Center for Biological and Environmental Nanotechnology (CBEN) discovered that superacids, such as sulfuric acid, work well in dispersing SWNTs into an easily processed form. This carbon form ranges from individually dissolved nanotubes to a liquid crystal that acts as a starting material for aligned SWNT fibers. This improved material, then, has laid the groundwork for larger objects made entirely of SWNTs. Figure 8-1 shows self-assembled carbon nanotubes.

Since carbon nanotubes are stronger than steel, they could help create super tough, nanotube-reinforced plastics. Use of SWNTs would cut the weight of airplanes, spacecraft, and ground vehicles considerably. The United States military is interested in using nanotubes for bettering new radar-absorbing coatings for tanks and personnel vehicles as well as stealth ships and planes. In aerospace design, the importance of SWNTs may be nearly as big a step as going from prop planes to jets.

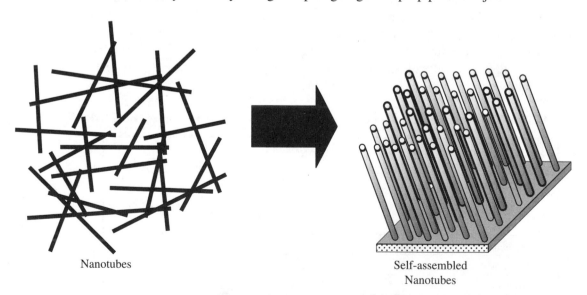

Nanotubes

Self-assembled
Nanotubes

Figure 8-1   Self-assembled nanotubes.

Multi-walled carbon nanotubes (MWNTs) with an average diameter of about 40 nm also have a variety of potential uses in everything from cell phone lens systems and shutter materials to car windows and sporting goods. Their compressive strength appears even greater than SWNTs and is proving important in composite materials.

Manufacturing

Tough technical puzzles and fabrication obstacles aren't slowing nanomaterial research. While in the short term, the greatest research efforts and funding are for high-performance materials such as nanowires or semiconductors, the nanotube manufacturing world is also expanding.

Some people think nanotubes will be most useful in applications such as coatings and paints. When mixed with paint, nanotubes become electrostatic. This helps paint and other nanotube-containing coatings stick more tightly to surfaces. Cars on the production line could be coated with nanotube pigments to cut manufacturing costs. Nanotube pastes might be able to improve liquid crystal and flexible displays to create images sharper than silicon and carbon-based films can generate.

Nanocrystalline Materials

As you know, everything is made up of atoms and molecules. Most bulk materials have particles varying in size from hundreds of microns (millionths of a meter) to a few millimeters.

Nanocrystalline materials have particles of around 1 to 100 nm. An average atom is about 1 to 2 angstroms (Å) in radius. One nanometer is roughly 10 Å, and there may be three to five atoms in one nanometer, depending on the size of the atoms.

Nanomaterials are particularly strong; hard; ductile (bendy/stretchy) at high temperatures; wear, erosion, and corrosion resistant; and chemically reactive. Many nanomaterials are much more useful than their commercially available bulk counterparts, depending on their various specific properties. For example, nanosilver has special catalytic properties that bulk silver doesn't have (e.g. interacting with and killing viruses).

The following five methods are commonly used to produce nanomaterials:

- Sol-gel (colloidal) synthesis
- Inert gas condensation
- Mechanical alloying or high-energy ball milling
- Plasma synthesis
- Electrodeposition

Although all of these processes are used to create various amounts of nanomaterials, currently sol-gel synthesis is able to

- make precision materials in large quantities fairly cheaply
- create two or more materials at the same time
- make extremely homogeneous (same throughout) alloys/composites and ultra-high purity (99.99 percent) materials
- produce materials (ceramics and metals) at ultra-low temperatures (around 150 to 600°F compared to 2500 to 6500°F in standard methods)
- fine-tune atomic composition/structure accurately

By creating or augmenting materials at the nanoscale, applications engineers can add capabilities such as superior strength to existing products.

Nanocrystals

Nanocrystals are clumps of atoms that form a cluster. They are bigger than molecules (~ 10 nm in diameter), but not as big as bulk matter. Although nanocrystals' physical and chemical characteristics change, one of their big advantages over larger materials is that their size and surface can be precisely controlled and properties tuned like *quantum dots* (a type of nanocrystal). In fact, scientists can tune how a nanocrystal conducts charge, decipher its crystalline structure, and even change its melting point.

Chemist Paul Alivisatos of the University of California at Berkeley and Lawrence Berkeley National Laboratory is making nanocrystals by adding semiconductor powders to soap-like films called *surfactants*. His group has grown mixtures of crystals using different surfactants. These react with semiconductor powders and produce different-shaped nanocrystals (e.g., rods instead of spheres).

> A *surfactant* is a substance (e.g., detergent) added to a liquid that increases its spreading properties by reducing its surface tension.

Alivisatos's knack for growing semiconductor nanocrystals in the shape of 2D rods opens up many new applications and shows how controlling crystal growth is important to changing size and shape. Although shape change is not completely understood, it's possible that the interaction of atoms with different surfactants causes a crystal to grow in a particular way. To keep up with a speedy growth rate, then (with the right mix of surfactants), crystals take on elongated, rod-like and faceted shapes that make the most of surface area. Figure 8-2 illustrates these nanocrystals.

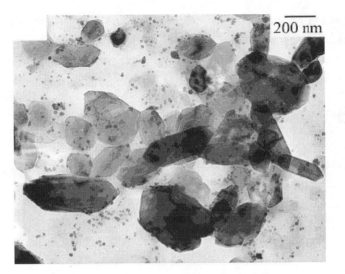

Figure 8-2 Nanocrystalline quartz particles.

Additionally, Alivisatos's group has shown that rod-shaped nanocrystals give off polarized light along their long axis compared to nonpolarized light fluoresced by cadmium selenide nanocrystal spheres. This is important in biological-tagging where marker attachment is a big deal.

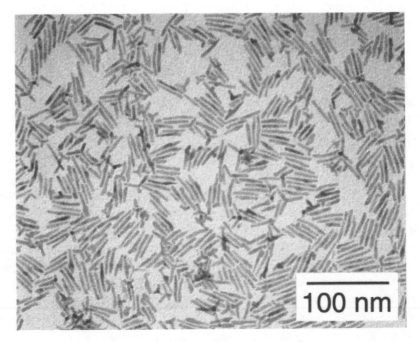

Figure 8-3 Formation of nanorods.

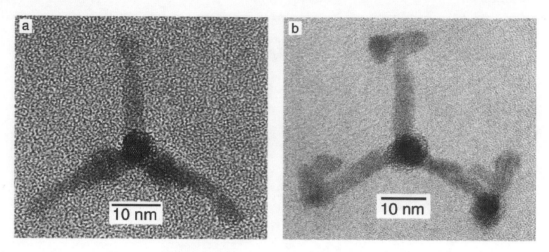

Figure 8-4   Tetrapod nanocrystals.

As far as other applications, Alivisatos's work has shown that the gap between emission and absorption energies is higher for nanocrystal rods than for nanocrystal spheres. This application would improve light-emitting diodes (LEDs) where light reabsorption has historically been a drawback. Since nanorods can be packed and aligned (like logs on a truck, see Figure 8-3) they may also work well in LEDs and photovoltaic cells.

Additionally, Alivisatos and others can change nanocrystal growth conditions and rates to create nanocrystals in the shape of teardrops, arrowheads, and even branched "jacks" shapes. Figure 8-4 illustrates a jacks form. Although these shapes don't have obvious uses right now, they could be important in the future. For example, since *tetrapod* (shaped like a jack) nanocrystals always land on three arms with the fourth arm extended, they have been suggested as possible connectors for innovative nanoelectronics.

QUANTUM DOTS

Quantum dots are semiconducting nanoparticles that are able to trap electrons in small spaces. They contain a single unit of charge and give off different colors of light depending on size and specific energy levels. These energy levels can be limited by changing the size, shape, and charge potential. Energy spacings and color are related (i.e, they appear to have a certain color because the energy emitted has an associated wavelength in the visible region of the spectrum). Changing the quantum dot size changes energy spacing and in turn affects a solution's visible color. Color change related to changes in particle size is a unique part of the nanoscale world.

Similar to quantum dots are *quantum wires* and *quantum wells*. Quantum dots confine electrons to negligible dimensions, compared to quantum wires (line) and quantum wells (flat) that have an area about the same as a de Broglie wavelength.

> A *de Broglie wavelength* is the measure of wave movement (wavelength) of a particle. The wavelength (λ) is given by $\lambda = h/mv$ (where h is the Planck constant, m is the particle mass, and v is its velocity).

This confinement allows energy levels and the measurement of electrical charge. Quantum dots are also important for optical applications because of their high *quantum yield* (i.e., the number of defined interactions that take place as a photon is absorbed in a system). They may also serve as components of a *qubit* (like a computer bit except at the nanoscale) for quantum information processing.

> A *quantum dot* (i.e., nanodot or qdot) is an inorganic, semiconductor nanocrystal that is less than or equal to 10 nm in overall size.

Like atoms, quantum dot energy levels can be studied with advanced spectroscopy methods. Because of their optical features, different sized quantum dots (with different color/wavelength absorptions) are not visible individually to the naked eye, but the color of their solution can be easily seen.

As we have learned, although quantum dot composition is important, color is related to size. The larger the dot, the more its fluorescence appears toward the red end of the light wavelength spectrum. The smaller the dot, the more blue it appears. This ties back to energy spacing effects described previously in the chapter. Additionally, some researchers believe quantum dot shape also affects colorization, but more work is needed to be sure.

Biological Markers

Another advanced quantum dot application includes the potential use of artificial *fluorophores* (fluorescent markers) for tumor detection using fluorescence spectroscopy during surgery.

We've learned that it is possible to *tune* a quantum dot. Size plays a key role. The larger and more red-shifted (in the light spectrum) a quantum dot becomes, the less quantum properties come into play. The smaller a quantum dot, the easier (relatively speaking) it is to take advantage of quantum effects.

Clinical researchers and physicians use quantum dot tuning characteristics in biological labeling as *markers* (e.g., antibodies attached to specific proteins). When quantum dots are taken up through a cell membrane, they are able to tag different parts within the cell (the enhanced permeability and retention effects discussed in Chapter 6). Better still, quantum dots don't seem to bleach (lose their color brightness) over time. This improves fluorescence imaging as researchers try to crack cellular mysteries.

Currently, various organic dyes are used in biological analyses. Eventually, as more complex imaging techniques become available, traditional biological dyes will become less useful. Quantum dots are superior to traditional organic dyes on several counts, including brightness (due to high quantum yield) and stability.

Researchers are studying spherical nanocrystals with structures of cadmium sulfide and cadmium selenide cores. These nanocrystals (depending on size) might be used to emit multiple colors of light. These may also be used in several different applications, including super sensitive fluorescent labels for use in studying biological materials. In fluorescent labeling, markers are tagged with dye molecules that fluoresce or emit a particular color of light when stimulated by photons from a confocal microscope.

Nanocrystals also offer possibilities as intravascular probes/sensors for imaging as well as drug delivery. Bioengineering researchers Maria Akerman, Warren Chan, Erkki Ruoslahti, and others at the University of California at San Diego are exploring the use of quantum dots for *in vivo* targeting. Their work has shown that stable and tunable nanocrystals, coated with a lung-targeting peptide, can be used to target mouse lung tissue. Additionally, two other peptides caused nanocrystals to move toward tumor blood and lymphatic vessels as well. These findings make directed medical treatments using nanoparticles a real possibility.

To study some cell populations, clinicians also need to look at combinations of markers. Some measurements need a multiple-color light release that is tough to get with regular dye molecules. Nanotechnology-based markers might take care of this problem completely.

It's apparent that nano-enhanced imaging advances have a good chance of bringing about real progress in medical diagnosis and treatment. Ultraviolet light-emitting devices could be attached to the end of a fiber-optic probe, for example, to treat damaged internal organs with direct UV irradiation. Other UV gadgets could be used in environmental applications such as reactors that destroy contaminating organisms or toxic organic waste.

ANALYSIS TOOLS

Analysis instruments are also important in obtaining nanomaterials of precise properties (e.g., computer components). Although nanocrystal arrays have been studied with atomic-force microscopes that basically feel their way across surfaces (refer back to Chapter 4), other methods are also in use.

Michael O'Keefe and Christian Kisielowski of the National Center for Electron Microscopy (NCEM, a U.S. Department of Energy facility where researchers carry out electron beam material characterization) developed new methods that are highly sensitive. NCEM's One-Ångstrom Microscope (OÅM) has achieved the United States' highest nanomaterial resolution of 0.8 angstrom (≤0.1 nm).

The OÅM's resolution can bring the different parts of many crystalline materials into sharp 3D focus. Additionally, the OÅM clarifies poor images normally present in a regular microscope resolution. Extra information can be deciphered with a computer by combining different images of the same sample. In other words, it can bring even fuzzy parts into focus.

Other new image-processing techniques make it possible for individual atoms lined up in a row to be counted. According to Kisielowski, this means theory and experiments are merging, so predictions of atom cluster size can be reliably confirmed. In fact, computer simulated and modeled nanomaterials are acting very much like observed atom clusters.

With new techniques and instrumentation, scientists can account for nearly every atom in a surface nanocluster. For example, columns of silicon atoms at a gate interface (the most important interface in integrated circuit technology) can be located with unmatched precision of about 0.01 angstrom.

QUANTUM COMPUTING

Quantum dots are a big technological leap for solid-state quantum computing. By applying small voltages to electrode leads, electron flow through quantum dots can be controlled. Good control helps designers make precise measurements of electron spin and other related properties. Data storage and calculations using quantum capabilities may soon become a reality.

Unlike electronic devices that use electron charges to carry signals, quantum computing uses electron spin or light polarization. This technique could greatly speed up operations, while lowering power loss in computing systems.

When a layer of crystalline zinc oxide (~10 atomic layers thick) is used between two sheets of zinc/manganese oxide (~500 atomic layers thick) and the two outermost sheets are attached to a battery forming a circuit, polarized charges are sent into the ultrathin sheet of zinc oxide. Electrons and holes (e.g., missing electrons) get back together in the middle layer. Holes with one spin type will interact only with electrons that have the same spin type. Consequently, polarized UV light is given off that may be used in optical quantum computing.

Along with being nanoscale small, quantum dots have better transport and optical properties and are good candidates for use in amplifiers, diode lasers, and biological sensors (e.g., within single cells).

Quantum dots have also appeared in many games and electronics. The latest technology in gaming and DVD players all use a blue laser for data reading. Blue lasers, once thought impossible, were developed through the creation of a quantum dot that emits blue light.

Nanotechnology and its impact on computing, electronics, sensors, and communications will be described more fully in Chapters 9 and 10.

Alloys

Materials scientists and engineers have long known how to make strong alloys, even before they understood a materials' structure. First the different components were heated until they dissolved together. Then, the alloy was allowed to cure for several days when one or more of the components would fall out of solution and create a much stronger, harder alloy.

In fact, *precipitation hardening* has been crucial in aviation since the Wright brothers' first powered flight. Modern testing has shown that the Wrights' engine was made from precipitation-hardened aluminum/copper seven years before such alloys were commonly used in aviation.

> An **alloy** is composed of two or more metals (e.g., brass is an alloy of copper and zinc) or a metal and a non-metal (e.g., steel is an alloy of iron and carbon).

In general, the closer the material spacing, the harder the alloy. Some additives to advanced alloys such as aluminum 2219 (copper and magnesium added) are so tiny and make up such a small percentage of the total alloy composition that the crystalline form and arrangement are not completely known.

Although these precipitates are only a few nanometers in size, their unique crystal structure, shape, and the way they strengthen an alloy can be found with transmission electron microscopy and an overall computer analysis.

Because precipitates change a material's crystalline structure, they can also affect *shearing* (the way something breaks). This is a problem when engineers depend on a materials' properties to perform in a certain way for a specific application. When the original material is made into a different structural alloy, shearing may be changed. Generally the thicker the alloy, the shorter the shear that can take place.

As alloy inclusions are incorporated at the nanoscale, thermal properties can also change. Energy between newly mixed materials becomes more important than each material's internal energy. So a nanoscale material may melt at much greater or lower temperatures than the original bulk material.

Applying these nanoscale characteristics, researchers at the NCEM patented a new alloy of aluminum, copper, germanium, and silicon for use in the aerospace and automotive industries. The new alloy has an extremely dense distribution of ultrafine precipitates with two different compositions. Harder and more stable than aluminum 2219, the new alloy is also more energy efficient.

By probing the atomic structure of nano additives to solids with electron microscopy, researchers are getting a lot of information on how materials can be controlled on the nanoscale.

Nanocomposites

The materials and processes used to separate nanoscale particles in plastics, metals, or ceramics is a big part of *nanocomposite* technology. The biggest use of these new nanocomposites is that they can be used to make ultrafine-grained structures with dimensions in the near nanoscale range, creating greater strength from much greater surface area.

Some clay nanoparticles are combined with other materials to form nanocomposites. These nano-sized clay particles are made up of *montmorillonite* (a soft silicate clay that expands when it absorbs liquids) and used to color paper and cosmetics. These types of modified clays are also used to make nanocomposites.

> *Nanocomposites* are a new class of materials created from a highly refined form of nanoclay that is mixed into plastic/ceramic resins.

Advanced nanotechnology tools and techniques, along with the much improved ability to understand, manipulate, and activate single molecules and nanostructures, allow material scientists a world of possibilities. Ever more important discoveries will be made with hybrid or nanocomposite materials (i.e., combining very different organic, inorganic, and biological systems in one integrated structure).

Nanocomposites also use clay nanoparticles to boost hardness. These applications have been used in automotive panels and van step bars.

Nanomaterials and NASCAR

NASCAR drivers often get burns from cockpit heat that comes through the engine's fire-wall, transmission, and floor. Nanomaterials have been used to cool extreme temperatures as high as 150°F inside speeding race cars.

When NASCAR racer Bobby Allison visited the Kennedy Space Center in Florida and learned how the space shuttle was protected from the excess heat of atmospheric re-entry by a special material, he wondered whether this technology would work in race cars, too. He called racer Roger Penske to discuss the possibilities.

Then, along with Rockwell and NASA, Penske's crew fit the NASA composite material to their No. 2 Ford Thunderbird stock car. The material added less than four pounds to the car's total weight and didn't change handling or balance. After several test track laps at top race speeds of around 200 mph, areas where the nanomaterial was added to the driver's cockpit showed temperature drops of more than 50°F.

If a lightweight, heat-absorbing nanomaterial were used for NASCAR and other formula cars, the racing field would be safer for both man and machine. Drivers would stay cooler and engines would function better without metal fatigue caused by extreme heat. Additionally, this application could offer benefits for other high temperature vehicles and applications, including firefighters' protective suits. Smart nanomaterials in extravehicular (EVA) suits keep astronauts protected from extreme heat and cold on spacewalks.

Nanorings

Researchers at the National Institute of Standards and Technology (NIST) labs have shown that when a gold ring is designed with a radius of about 60 nm, it takes on special properties that make it a useful container for nanoscale experiments. These experiments are important to biochemists and the drug industry in perfecting new medicines.

Metal nanoparticles are important because they absorb and emit certain types of light very efficiently, depending on particle size and shape. Recently, NIST physicists and collaborators in Sweden and Spain documented that gold *nanorings* have unique optical and electromagnetic properties that can be tuned by changing the ratio between the ring's radius and wall thickness.

> When light is shown on a ***gold nanoring***, a strong electromagnetic vibrating field (in the near infrared) is created inside and around the ring.

When light is aimed at the ring, it excites the metal's electrons and creates a wave that oscillates in certain ways depending on the light's wavelength and geometry. Imagine wind hitting a small pond, where wave motion depends on wind speed and pond size and shape.

The NIST team discovered how to synchronize the electron pool and incoming light's energy so that they vibrate at the same wavelength.

The vibration they observed caused a strong, electromagnetic field to oscillate inside the ring cavity. Figure 8-5 shows the field within and around a ring with a radius of 60 nm and a wall thickness of 10 nm. The arrows indicate the direction of the field created by the pool of excited electrons.

The electromagnetic field inside the ring works in the near infrared part of the electromagnetic spectrum. The ring cavity can then be used as a container for testing molecules with light that intensifies infrared signals. For example, as discussed in Chapter 6, researchers study biosystem proteins and chemical reactions by treating them with lasers and recording how much light is absorbed and re-emitted at

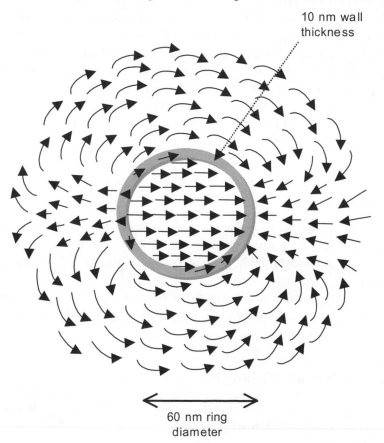

Figure 8-5   Within a nanoring, light creates a strong electromagnetic vibrating field.

specific frequencies. By doing these types of experiments inside a nanoring, researchers can get better infrared signals and clearer information.

SELF-ASSEMBLY

Associate professor of chemistry in Purdue's School of Science, Alexander Wei and his research team have created nanorings from cobalt nanoparticles. Less than 100 nanometers across, the nanorings store magnetic information at room temperature. Most importantly, these nanorings can also form by *self-assembly*.

Cobalt nanoparticles link up and form into rings that act like tiny magnets with a north and south pole when they get near each other. The magnetism in nanoring formation also creates a collective magnetic condition known as *flux closure*. Although there is strong magnetic force (i.e., flux) within the rings themselves that comes from the particle's magnetic poles, the total magnetic effect outside is zero after the particles form rings.

Wei's research indicates that magnetic rings may serve as memory elements in devices for long-term data storage and RAM (random-access memory). The rings have a magnetic field that rotates backward or forward like the hands of a clock, storing binary information. Early data has shown that nanorings' magnetic states can be controlled. When applying a magnetic field, a nanoring "bit" can be switched back and forth between 1 and 0. Then if nanorings can be tied to conductive nanowires, data storage applications will abound.

Nanocoatings

Nanocoatings on tennis balls create better seals that keep them from losing air. This makes the standard tennis ball last about six times longer than balls using other sealants, before it loses its "new" bounce. Increased savings are a big plus of nanocoatings as well.

Nanoclays and nanoparticle coatings are being used in everything from tennis balls and bikes to cars. They improve bounce, strengthen high-impact parts, and make surfaces scratch-proof.

DEFENSE AND AEROSPACE

Coating processes have a lot of applications in the aerospace and defense industry as well. Some of these include the following:

- improving the durability, reliability, and performance of various components

- erosion, sliding, and general wear resistance
- improving surface quality
- corrosion resistance against pitting, peeling, oxidation and heat

In fact, several multi-functional nanocoatings are being created for use in aerospace applications, which will provide corrosion protection using environmentally safe materials. They are also expected to be able to detect corrosion and mechanical damage to aircraft skin; react to chemical and physical damage, improve adhesion, and make metals parts last longer. Lightweight, high-strength, heat stable nanomaterials are also being studied for aircraft engines.

In the defense industry, conventional paints are labor intensive to apply and potentially hazardous to the people working with them. These coatings have to be touched-up by hand, which can hide metal damage. As a result, the total cost for U.S. Department of Defense corrosion-related problems is $10 billion per year ($2 billion in painting and paint-scraping operations). Smart coatings will make it possible for military vehicles, if corroded or scratched, to detect and heal their surfaces. There is also the long term potential for vehicles to change color on the battlefield, creating instant camouflage and making tanks and other military vehicles harder to detect.

Nanoshells

Although we discussed nanoshells in Chapter 6, they are an important nanomaterial for smart materials. Nanoshells are a new type of optically tunable nanoparticle made from a *dielectric* (e.g., silica) core coated with an ultra-thin metallic (e.g., gold) outer layer. Gold nanoshells like gold colloids have a strong optical absorption from the metal's electronic response to light. For example, the optical absorption of gold colloids produces the bright red color used in home pregnancy tests.

By comparison, gold nanoshells' optical response depends on the different sizes of the nanoparticle core and gold shell thickness. Like quantum dots and gold nanorings, size affects optical tuning properties. By changing proportional core and shell thicknesses, gold nanoshells can change color across visible and near-infrared light spectrums. Gold nanoshells can also be made to either absorb or scatter light by changing the particle size in relation to optical resonance and wavelength.

Nanoshells, originally, discovered by Rice University Professor Naomi Halas, combine chemistry (nanoshell fabrication), physics (optics), and engineering (fabrication fine-tuning). Joined by Rice bioengineer and research colleague, Jennifer West, Halas and West founded Nanospectra Biosciences, Inc. in 2001 to develop nanoshells for medical applications. Figure 8-6 shows a gold nanoshell surrounded by polyacrylamide gel and antibodies that can attach specifically to a tumor's surface.

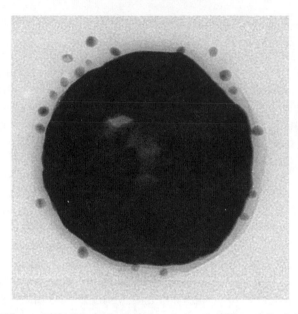

Figure 8-6   Nanoshell with polyacrylamide gel "halo".

The many possibilities offered by nanoshells are currently being tested by researchers at Rice University in the following areas:

- Whole blood optical immunoassays
- Optical imaging contrast agents
- Photothermal ablation (cooking) of cancers and macular degeneration (a medical condition where light sensing cells in the eye stop working over time)
- Pharmaceutical delivery
- Optically controlled microfluidics valves
- Biosensing

Additionally, nanoshells are useful in applications like inhibition of photo-oxidation of polymer films.

Catalysts

For years, the nanotechnology field has been plagued by sci-fi ideas of tiny robotic, molecular assemblers scurrying around cramming atoms together. But natural molecular assemblers have existed for a long time in the form of *catalysts*.

> *Catalysts* are substances that increase chemical reaction rates without being consumed or undergoing any permanent chemical change.

Nature's catalysts, *enzymes*, assemble specific end products. They provide an alternative pathway by which reactions take place at lower activation energies (i.e., energy needed for a reaction to take place). Industrial catalysts are not so precise. They are often made of metal particles on an oxide surface, working in hot reactant streams. The smaller the catalyst particles, the greater the ratio of surface to volume (i.e., the more catalyst surfaces are exposed, the greater the reaction efficiency).

As far back as the 1920s, industries began putting metal particles on to support surfaces by precipitating salts out of solution. However, these particles were of different sizes with random spacing.

Size and spacing have been found to be critical in making industrial catalysts with the efficiency and precision of natural enzymes. Testing has shown that single platinum crystals around 15–20 nm high and 100 nm apart can be grown on a silicon oxide surface about 0.5 cm^2. In fact, even if the surface area is halved, platinum cluster arrays can be more than 20 times more active than platinum alone. The interface between the metal and the oxide was important in catalysis and could lead to the development of super efficient nanocatalysts.

Nanotechnology has great potential to expand catalyst design for the chemical, petroleum, automotive, pharmaceutical, and food industries among others. Specialized nanocatalysts that interact with biological structures may serve as important bridges between traditional and enzymatic catalysis.

Microcapsules

Microcapsule research is one of the most active areas in materials nanotechnology. Many companies are developing hollow containers for everything from drug delivery and imaging markers to sunscreens, cosmetics, and perfumes. This ability to encapsulate (enclose) molecular payloads is important to new medical and industrial applications.

> *Microcapsules* self-assemble at room temperature when nanosilica particles are mixed with a solution of polymer and salt in water.

Microcapsules are also able to encapsulate enzymes, the complex biomolecules that start and/or control many cellular processes. Dr. Michael Wong, assistant professor of chemical and biomolecular engineering and chemistry at Rice University, has shown how enzymes can be stored inside microcapsules without leaking through the walls, although smaller molecules were able to move through the microcapsule structure. Figure 8-7 shows the simplicity of these spheres.

Since this capability allows enzymes to catalyze chemical reactions with other molecules, Wong believes that the technology may be useful as *microbioreactors* for industrial chemical and pharmaceutical production lines.

Unique properties of nanomaterials and nanoscale structures have triggered an avalanche of materials research. Improved product performance using nanomaterials (e.g., fillers in plastics, surface coatings, and computer circuits) is already taking place. Nanotechnology holds great promise for the future and has every indication of leading to even more changes in products, processes, and public awareness.

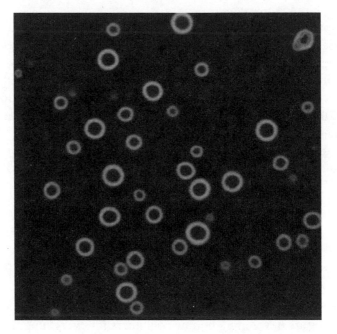

Figure 8-7 Microcapsules.

Quiz

1. The measure of wave movement (wavelength) of a particle is known as

   (a) a span

   (b) milliliter

   (c) light year

   (d) de Broglie wavelength

2. Gold nanoshells' optical response depends on the

   (a) amount of heat used during fabrication

   (b) current market value of gold

   (c) different sizes of the nanoparticle core and shell thickness

   (d) time of day that testing is performed

3. A new class of plastics created from a highly refined form of nanoclay mixed into plastic resin is known as a

   (a) nanocomposite

   (b) nanoring

   (c) nanotube

   (d) nanostyro

4. Catalysts are substances that

   (a) cause earthquakes

   (b) increase reaction rates without being consumed or undergoing any permanent chemical change

   (c) are found in marshy areas

   (d) harden after processing all reactions

5. Microcapsules are able to encapsulate

   (a) sunlight

   (b) quail eggs

   (c) electrical energy

   (d) enzymes

6. When nanomaterial thermal coatings were added to a stock car cockpit, internal racing temperatures dropped by more than

  (a) 15°F

  (b) 35°F

  (c) 50°F

  (d) 60°F

7. Nanorings' unique optical and electromagnetic properties can be tuned by changing the

  (a) ratio between the ring radius and wall thickness

  (b) hand the ring is worn on

  (c) rings' diameter to approximately 3 mm across

  (d) pressure under which they are formed

8. Nanoshell development combines chemistry, physics, and

  (a) anthropology

  (b) music

  (c) engineering

  (d) ecology

9. Spheres that self-assemble at room temperature when nanosilica particles are mixed with a solution of polymer and salt in water are called

  (a) tapioca

  (b) buckyballs

  (c) microcapsules

  (d) billiards

10. Nanocrystals are made by injecting semiconductor powders into hot, soap-like films called

  (a) ideal gases

  (b) surfactants

  (c) enzymes

  (d) mousse

CHAPTER 9

Electronics and Sensors

In Texas, bigger is better. Home of longhorn cattle and ten-gallon hats, anything smaller might as well not bother to show up. However, when it comes to electronics, Texans, like everyone else, want smaller, more capable, and faster computer components. If technology is to keep meeting the public's computing needs, innovation must be in the forefront.

Moore's Law

Three years before co-founding Intel, Gordon Moore noted that the computer industry was doubling the density of components every year. He said that if this continued every year, the present chip density of 50 components per circuit might reach 65,000 components per circuit by 1975. This speculative guess became known as

Moore's Law and since then has served as the operational standard for the semiconductor industry. Figure 9-1 illustrates Moore's Law and current trends.

Besides the technical progress described by Moore's Law, there is also an economic angle. Like every industry, factory costs for building electronic components nearly double annually. In 2005, component size was in the 130 nm range and diving, with millions of components per circuit. Intel has plans to construct a factory that will produce features as small as 65 nm. At this size, components nearly 300 atoms wide would be possible.

This ever-shrinking component size trend doesn't appear to be slowing down, but it will eventually hit a wall. At some point, the smallest component will be about the size of an atom. If that occurs, and it seems likely that it will, computing will have to make a quantum leap in how information is transferred. Some people think this will take place by 2010, while others say 2020 to 2030 is more likely. After components reach the atomic level, the next step may be computers with working parts made of subatomic particles. Engineers may be able to get past component size limitations by using nanostructures in new ways (e.g., by using nanotubes as wires).

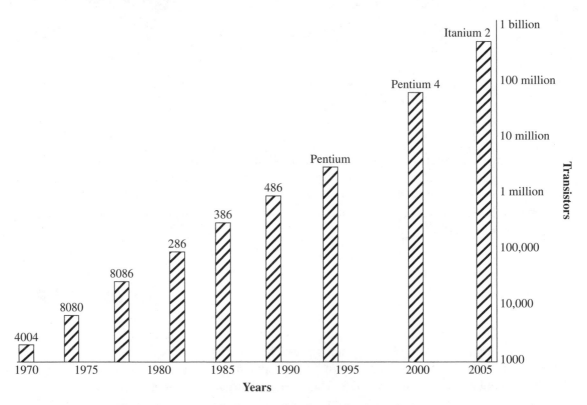

Figure 9-1   Moore's Law predicted growth of circuit components.

What about software? Will super fast computers have to write software for other super fast computers?

QUANTUM QUESTIONS

Nanotechnology's ability to construct bottom-up materials atom-by-atom is not just about making things tinier. The semiconductor industry needs smaller tools and components to continue manufacturing new generations of chips and to keep and create jobs.

Nanotechnology moves into the "really cool" realm when matter is reduced down to the molecular level. As described in Chapter 3, breaking a material down into nanoparticles multiplies its surface area by a factor of millions. Greater surface area equals higher reactivity. Nanomaterials melt, ignite, and/or absorb much more quickly. Gold, for example, formed into the shape of a brick, ring, or tooth, is *inert*. Nanogold, however, can function as a powerful catalyst.

> An *inert* material has few or no active properties and doesn't react with things around it.

Some nanomaterials, when scaled down smaller than a wavelength of visible light, become transparent. This makes it possible for scientists and engineers to make opaque materials, such as silicon, able to transmit light. Other nanomaterials get stronger as they get smaller. Carbon nanotubes with an atomic structure similar to diamond become tough structures that combine strength and flexibility.

Much about the nano world is still unknown. For one thing, nanomaterials don't follow Newtonian physics (i.e., gravity). At the large scale, the laws of gravity, optics, and molecular movement are measured as averages, not the zippy actions of a single nanoparticle. To understand an individual nanoparticle's properties, researchers must explore *quantum theory*.

> *Quantum theory* describes how energy is not absorbed or given off constantly, but in spurts and only in multiples of specific, nondivisible energy units called *quanta*.

Quantum mechanics (a branch of physics) deals with the behavior of matter at the atomic, nuclear, and particulate levels. At these levels, energy, momentum, mass, and other properties don't change constantly, as they do at the bulk level. They come in specific units, or *quanta*.

While Newtonian physics can easily explain the movement of the planets and the arc of a baseball hit out to center field, quantum mechanics is thought to do a better job of describing molecular occurrences. This includes systems' behavior at the nanoscale and smaller. In fact, Newtonian mechanics has trouble explaining the existence of stable atoms as well as larger systems such as superconductors and superfluids (advanced lubricants).

Quantum mechanics' predictions have held true for more than a century's worth of experiments. Quantum mechanics cover at least three types of molecular activity that regular physics doesn't quite explain. These include the measurement of individual physical quantities; *wave particle duality* (i.e., wave and particle like properties by a single particle); and *quantum entanglement* (how measurements in one system influence other systems at the same time).

Quantum theory is the backbone of electronics and semiconductor research. When quantum physics is better understood, companies that navigate this territory successfully will be the powerhouses of the nano era.

Transistors

In 1948, American physicists John Bardeen, Walter H. Brattain, and William Shockley, working at Bell Telephone Laboratories, announced the invention of the transistor. (In fact, they were jointly awarded a Nobel Prize in 1956 for the invention. Independently, the transistor was also created at almost the same time by Herbert Mataré and Heinrich Welker, German physicists at Westinghouse Laboratory in Paris.) Since that time, many different transistor types have been designed.

Bardeen demonstrated his single-mindedness in this complex scientific area when he was again awarded the Nobel Prize with Leon N. Cooper and John R. Schrieffer for the development of the superconductivity theory in 1972. As of this writing, Bardeen is the only person to win two Nobel Prizes in the same field.

To understand nanoelectronics, you must first understand the fundamentals of how computers work. The *transistor* has replaced the vacuum tube as the main electronic signal regulator. Transistors are very small, long-lasting, resistant to physical shock, and fairly cheap to make. Originally, individual devices were enclosed in a ceramic case, with a wire extending from each side. The transistor was connected to an electrical circuit. Although single, isolated transistors are still used, most of today's transistors are built as components of integrated circuits. Transistors are found in nearly in all electronic devices (computers, radios, global positioning receivers, space craft, and guided missiles).

> A *transistor* regulates current/voltage flow and acts as a switch or gate for electronic signals.

A transistor is made up of three layers of sandwiched semiconductor material that is able to carry a current. A *semiconductor* is a material such as silicon or germanium that conducts electricity in an ordered way. Halfway between a true metal conductor, such as copper, and an insulator, such as rubber, a semiconductor has the properties of both. At high temperatures, semiconductors are conductive like metals, and at low temperatures they act like insulators. In a semiconductor, the movement of electrons is limited, depending upon the crystal structure of the material used.

> A *semiconductor* is a material with an electrical conductivity halfway between those of an insulator and a conductor (i.e., an insulator at low temperature and a conductor at room temperature).

In solid-state electronics, either pure silicon or germanium is used as the baseline material for fabrication. Each has four *valence* electrons (electrons that are available for bonding), but at certain temperatures, germanium has more free electrons and a higher conductivity. Silicon, however, is the most universally used component of semiconductors, since it can be used at much higher temperatures than germanium. When charged with electricity or light, semiconductors change their state from nonconductor to conductor, or vice versa.

A semiconductor material is created by a chemical process called *doping*. Certain chemicals (impurities) are added to the starting material (e.g., silicon, germanium, and so on) to give it the benefits of both a conductor and an insulator. After doping, the finished material contains a small number of atoms with either more than four valence electrons (called an *n-type*, where *n* stands for negative due to the extra negative charges) or fewer than four valence electrons per atom (called a *p-type*, where *p* stands for positive due to the positive charge resulting from missing electrons) scattered throughout the material's crystal structure. The transistor's three-layer construction includes an *n*-type semiconductor layer sandwiched between *p*-type layers or a *p*-type layer between *n*-type layers. The affects of these types are discussed later in this chapter when we look at different transistor types.

A small current or voltage change at the inner semiconductor layer (acting as a control electrode) quickly causes a big change in the current passing through the whole component. The transistor acts as a switch, opening and closing an electronic gate super fast (dozens of times per second).

Currently, computers use circuitry made with complementary metal oxide semiconductor technology. These circuits use two balancing transistors per gate (e.g., one each of *n*-type and *p*-type materials). Transistors are the basis of integrated circuits, with huge numbers of transistors woven throughout and forming a *silicon chip*.

LITHOGRAPHY

Silicon and current microprocessor technology will soon reach its physical limit. Chipmakers will have to find other technologies that will allow them to fit more transistors onto silicon and build more powerful chips.

Lithography is a lot like photography in that it uses light to transfer images onto a substrate. While a camera uses film as a substrate, the substrate used for electronics is silicon. To create an integrated circuit pattern for a microprocessor, light is shown onto a *mask*, a sort of stencil with a specific circuit pattern. The light shines through the mask and then through a series of optical lenses that serve to shrink the image further. This smaller image is then projected onto a silicon, or semiconductor, wafer.

The wafer is layered with a light-sensitive, liquid plastic called *photoresist*. The mask is placed over the wafer, and when ultraviolet light (i.e., wavelengths of ~248 nm) shines through the mask and hits the silicon wafer, it hardens the portion of the photoresist that isn't protected by the mask. The covered photoresist (not exposed to light) stays gooey and is chemically washed off, leaving the hardened (cured) photoresist and exposed silicon wafer. Figure 9-2 shows how the process of etching a silicon surface works.

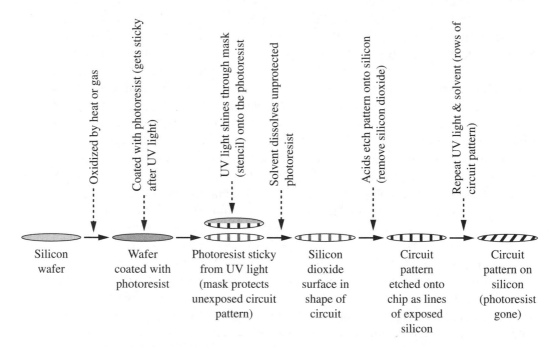

Figure 9-2 Making a computer chip using the photoresist method.

The process is repeated until the entire silicon surface is etched into the desired transistor pattern. When using light wavelengths of 248 nm, circuit widths of about 200 nm are achieved. Millions of transistors are used for the hundreds of microprocessors needed to operate at the top computer speeds required. That's why companies such as Intel have to keep making transistors smaller and smaller to fit more of them onto the wafer.

One factor in making microprocessors more powerful is the light's wavelength size: the shorter the wavelength, the more transistors that can be etched onto the silicon wafer. Then, the higher the number of transistors, the more powerful and faster a microprocessor will operate. That's why an Intel Pentium 4 processor with 42 million transistors is faster than the Pentium 3 with 28 million transistors.

Lithography, the process that replicates a circuit pattern chip to chip, wafer to wafer, or substrate to substrate, also controls the throughput and cost of electronic systems. A lithographic system includes an exposure tool, mask, resist, and processing steps to complete the pattern transfer from a mask to a resist and then onto integrated circuit devices. When any of these steps can be shortened or made more efficient, costs goes down.

> *Lithography* is the process of imprinting patterns on semiconductor materials for use in integrated circuits.

Today's silicon chips are packed tightly with transistors by *deep-ultraviolet lithography*, a photography-like method that focuses light through several lenses, carving integrated circuit patterns onto silicon wafers. This works to a point, but physics/light wavelengths limit how small chipmakers can go.

Using extreme-ultraviolet light (i.e., wavelengths about 10–15 nm) to carve transistors in silicon wafers, microprocessors up to 100 times faster than today's most powerful operating chips and memory chips with huge increases in storage capacity will be produced. Many chip manufacturers are already looking at *extreme-ultraviolet lithography* (EUVL) to lengthen silicon's usefulness until 2010.

Immersion Lithography

Immersion lithography is a variation that uses a liquid pool between the optics and the wafer surface, replacing the usual air gap. For a 193 nm UV wavelength process, ultra-pure, degassed water (no bubbles) is most often used as the immersion liquid.

Immersion lithography increases the effective depth-of-focus for a specific aperture and makes it possible to use optics with numerical apertures above 1.0. Immersion techniques may help augment better and faster computing.

TRANSISTOR TYPES

A transistor is made of up semiconductor materials that share the same area. Silicon, gallium-arsenide, and germanium are most often used as semiconductor materials with impurities added through doping. In *n*-type semiconductors, the added impurities (e.g., arsenic) cause an excess of electrons (negative charges), and in *p*-type semiconductors, the dopants (e.g., boron) cause a deficiency of electrons (extra positive charges) or "holes" in the silicon crystal lattice.

Junction Transistor

The *n-p-n junction transistor* is made up of two *n*-type semiconductors (i.e., emitter and collector) divided by a thin layer of *p*-type semiconductor (base). When the electric charges on transistor segments are correct, a small current between the base and emitter connections increases the current between the emitter and collector connections. This allows the current to be amplified. Figure 9-3 illustrates a simple transistor structure.

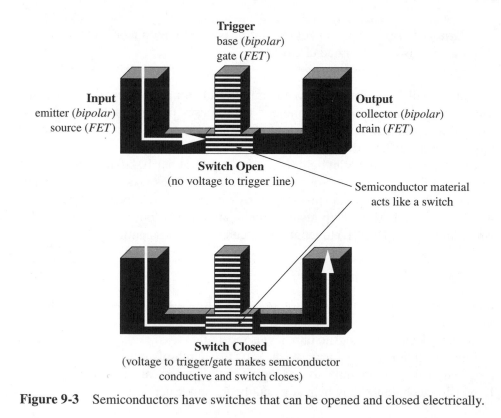

Figure 9-3   Semiconductors have switches that can be opened and closed electrically.

Circuits also use transistors as switching devices. Any current in the base-emitter junction generates a low-resistance path between the collector and emitter. The *p-n-p* junction transistor, stacked so that a thin *n*-type semiconductor layer is between two *p*-type semiconductors, works in the same way, except that all the polarities are reversed.

Field-Effect Transistor

Following the junction transistor, the *field-effect transistor* (FET) was created. Because it pulls almost no power from an input signal, it provides a huge advantage over the junction transistor. An *n*-channel FET is formed with a channel of *n*-type semiconductor material that goes between and contacts two small regions of *p*-type material near its center. The terminals attached to the ends of the *n*-type channel are called the *source* and the *drain*.

The terminals attached to the two p-type regions are known as gates. Voltage sent to the gates is routed so that no current goes across the junctions between the p- and n-type materials, so it's called reverse voltage. Differences in reverse voltage amounts bring about changes in the channel's resistance, permitting the reverse voltage to control the channel's current. A p-channel device works the same way, but with opposite polarities.

The metal-oxide semiconductor field-effect transistor is a type where a single gate is separated from the channel by a layer of metal oxide. This serves as an insulator. The gate's electric field extends through the dielectric and controls the channel's resistance. When applied to the gate, the device's input signal raises and lowers the current passing through the channel.

Nanotransistors

An innovative kind of *bipolar transistor*, created by scientists at the University of Illinois at Champaign-Urbana, has cracked the 600 gigahertz computer operating speed barrier. A terahertz transistor for high-speed computing/communications applications might now be possible. In addition to being smaller, the terahertz chip would have 25 times the number of transistors as the best Pentium 4 processor, would run 10 times faster, and would have no spike in power usage (a previous limitation). In fact, Intel plans to produce a terahertz transistor based on the new nanoscale transistor design advance.

Constructed from indium phosphide and gallium arsenide, a bipolar transistor is designed with a stacked composite collector, base, and emitter to lower transit time and increase current density. With a bipolar transistor, scientists have demonstrated a speed of 604 gigahertz, making it today's fastest operating transistor.

To get super high speeds in regular transistors, current density would run too high and melt the components. The new composite bipolar transistor operates at higher frequencies with lower current density. The new material's structure allows computer scientists to envision the likelihood of a terahertz transistor in the not-too-distant future. Faster transistors would then make computers faster, wireless communications more flexible, and electronic military systems more effective.

INTEL

So how does this development relate to nanotechnology? In the past 35 years, smaller and smaller components on shrinking circuits took Intel to the nano frontier and the 100 nm boundary. In fact, Intel plans to create components well within the nanoscale and to create transistors as small as 20 nm by 2007. This push would make Intel the world's largest nanotechnology company.

Thirty-five years ago, silicon memory cost almost 100 times more than the hand-woven magnetic cores that make up the memory components of today's computers. With the competitors' hounds at their heels, Intel's founders took a leap of faith and changed direction. They created the first microprocessor (Intel 4004) that encompassed 2300 transistors in a space a little larger than a thumbnail. The 4004 cost a whopping $200 per chip and processed an instruction every 1/60,000 of a second. More than 250,000 of the 4004 chips would be needed to match the performance of today's home PC processor. Since nothing fuels growth like success, Intel set to work putting even more functionality into generation after generation of ensuing microprocessors.

Electronics Competition

Most projections set the time it will take for the semiconductor industry to reach a physical size limit between 2010 and 2016 with standard methods. At that time, it will not be possible for engineers to shrink circuits any further onto tiny silicon wafers.

This short lead time will be significant for computer companies that want to continue to lead the pack in innovation, speed, and storage. Many industry watchers are wondering whether today's top dogs of the semiconductor industry can hang on to their leadership as they chart new paths for the industry as a whole. Many question whether new components and computing architecture will come from within or arise from nanotechnology.

Many semiconductor companies are thriving. They collected roughly $214 billion in revenue in 2005. It will be a long time before nanoelectronics start-up companies challenge that figure. However, to keep doubling computing capacity every 18 months, as Moore's Law describes, the computer industry must find new technologies, architectures, and materials. You wouldn't think that would be too tough since chipmakers have already manufactured chips at the nano level, with some chip nodes only 90 nm apart. Technically, these semiconductors are using nanotechnology.

Until the second decade of the twenty-first century, chipmakers will be dumping boat loads of money into nanotechnology research and will likely buy up nano startups (such as university spin-offs) and incorporate new nanotechnology methods and materials within existing chip manufacturing where possible.

DNA MOLECULAR COMPUTING

The molecules that make up our genetic code may form the basis of the computers of the future. *DNA chips* (also called *microarrays*) are part of a technology that offers applications in genetic research and diagnostics. DNA chips or arrays are devices in which different DNA sequences are arranged on a solid support (e.g., silicon, glass, plastic, and so on). DNA array technology may also play a future role in enabling nanofabrication.

DNA chips make it possible for a scientist to use thousands of probes at the same time. Various probes are lined up on a surface, with each sequence in a specific spot. The unknown sample is placed on the chip, and it will stick to its matching probe sequence on the surface. DNA arrays contain from 100 to 100,000 different DNA sites (pixels) on a chip's surface. Depending on the chip, the sites can range from 10 to more than 100 microns size. Each DNA site can contain from 106 to 109 DNA sequences.

Several companies are involved in developing DNA chips and arrays, including Affymetrix, PE Applied Biosystems, HySeq, Nanogen, Incyte, Molecular Dynamics, and Genometrix. Current DNA chip devices have applications in genomic research, pharmacogenetics, drug discovery, forensics, gene expression analysis, cancer detection, and infectious and genetic disease diagnostics.

More recent versions of electronically active DNA microarrays (developed by Nanogen) that produce controlled electric fields at each site may have potential applications for nanofabrication. These active microelectronic devices can transport charged molecules (e.g., DNA, RNA, proteins, and enzymes), nanostructures, cells and micron-scale structures to and from any test site on the device surface.

These DNA computing devices use electric fields to direct the self-assembly of DNA molecules at specified sites on the chip's surface. They also serve as semi-conductor motherboards for the assembly of DNA molecules into complex 3D structures. The DNA molecules themselves have programmable and self-assembly properties and can be customized with different molecular electronic or photonic capabilities.

DNA molecules can also be attached to larger nanostructures (e.g., metallic and organic particles, nanotubes, microstructures, and silicon surfaces). In the future, microelectronic arrays and DNA-modified components may allow scientists and engineers to direct self-assembly of 2D and 3D molecular electronic circuits and devices within larger semiconductor structures. In that case, electronically directed DNA self-assembly could encompass a broad area of potential applications.

For example, researchers in Israel have already tied transistors to strands of DNA. A team led by physicist Erez Braun at the Technion-Israel Institute of Technology developed a two-step process that did just that. First, they used binding proteins that made it possible for carbon nanotubes to bind to specific DNA strand sites. Then they formed the rest of the DNA molecule into a conducting wire. Since DNA (good at building things at the molecular level) does not conduct electricity, the research team needed to attach a metal conductor to the DNA.

To make it happen, Braun's team coated a main part of a DNA molecule with proteins from an *E. coli* bacterium. Then, carbon nanotubes (coated with antibodies) were added that stuck to the protein. Next, a solution of silver ions was added. The ions chemically bonded to the DNA's phosphate backbone, but only where no protein had attached. Aldehyde (carbon with a charged side group attached) then changed the ions into silver metal, creating the basis for a conducting wire.

Gold was added to coat the device and create a fully conducting wire. In this way, the carbon nanotube device was connected at both ends by a gold and silver wire and functioned as a transistor when varying voltage was applied across the substrate. Depending on the changing/applied voltage, the nanotubes either bridged the gap between the wires to complete the circuit or they did not. Using the new biological process, the team was able to connect two of the devices.

This same method may allow the manufacture of elaborate self-assembling DNA structures and circuitry. Although this is just a nano step towards DNA molecular computing, there is hope that in future decades, large scale, self-assembled electronic devices (e.g., computers) will be built and used widely. Then, operating at the atomic level, a scale far smaller than currently used, scientists and engineers will cause these super small molecular devices to transmit information, and in turn, transform computing.

What's the Hold Up

When chipmakers look for new materials for future integrated circuitry, single-walled carbon nanotubes (SWNTs) are a great choice. However, they are extremely tough to work with.

Measuring 1 to 5 nm in diameter (around 1/50,000 the width of a hair) and nearly 100 times stronger, 1/6 as heavy, and 20 percent more flexible than steel, carbon nanotubes are a dream material. They also have superior heat-retaining capacity, with almost no thermal leakage, and they can transmit an electrical charge at twice the speed of silicon embedded circuits. These strengths make them amazing choices for smaller, faster, cooler running chips. It has even been suggested that carbon nanotubes could lead to 3D integrated circuits, in which transistors are stacked not only side-by-side, but above and below each other—something not possible with silicon.

Uniformity

While nanotubes are "all the rage" with nanotechnology supporters, established companies want to get a handle on the material's potential. The Semiconductor Industry Association (SIA), a trade group whose membership includes major U.S. chip manufacturers, has joined with government agencies to begin a program to evaluate the potential of various nanoelectronic processes and promote government and private research funding.

The SIA will focus on two or three new nanotechnology areas, with the best products/processes getting the most funding. Semiconductor engineers are looking for low cost, dependability, and new design capabilities. The problem has been getting uniformly reliable nanotube growth.

Until 2005, most researchers weren't able to figure out how to make nanotubes in consistent, commercially available amounts. During growth, nanotubes form in different widths and lengths, and with various semiconductor and conductive properties. In fact, millions of carbon nanotubes (a sticky, knotted, microscopic jumble) look like a smudge of black soot. For researchers to get specific and regular types/sizes, they had to depend on a few high quality labs for samples.

Recently, however, great progress in growing nanotubes cleanly and consistently was made by researchers at DuPont, Rice University, and elsewhere. They were able to separate nanotubes by bonding non-attracting compounds to their walls.

"Imagine nanotubes are spaghetti and we've just invented butter," described the late Professor Richard Smalley, nanotube champion, Nobel laureate, and all around good guy.

In 2005, the uniformity issue faded. Researchers at Southwest Nanotechnologies grew carbon nanotubes of constant diameters, and collaborators at Duke University and Los Alamos National Laboratory grew the longest nanotubes to date (4 cm). With uniform and batch quantities of carbon nanotubes now possible, electronics applications and manufacturing can move ahead much faster.

NANOWIRES

As great as nanotubes' qualities are, fabrication methods that purify, untangle, straighten, and sort nanotubes are a lot more complex than growing large, silicon crystals. Building circuits with nanotubes is still a technical challenge that will keep scientists and engineers busy for quite a while.

> A *nanowire* has dimensions of a few nanometers (10^{-9} m). At this size, quantum mechanical effects are in play, so these wires are also known as *quantum wires.*

Much easier to work with, silicon *nanowires* (also called *nanorods* or *quantum wires*) are the next step down the road away from the silicon-only camp. Like nanotubes, nanowires form complicated configurations of super small transistors, but they don't have nanotubes' extreme strength.

Unlike carbon nanotubes, nanowires can form complex systems and architectures with various components that nanotubes don't easily work with. Silicon nanowires are a natural extension of present manufacturing technology. Additionally, engineers are able to control structure and maintain performance in ways that have been understood for many years.

Silicon, however is not the only material useful for nanowires. Depending on the application, metallic or multi-layered nanowires made from gold, copper, or manganese (to name a few) have incredible selectivity and specificity. With diameters down to 12 nm, nanowires can be used for optical, magnetic, sensing, solar cell, and/or electronic applications. Figure 9-4 lists the many choices available for nanowire fabrication.

Suspended Nanowires

Nanowires are made in a laboratory either through *suspension* or *deposition*. A suspended nanowire is created in a vacuum chamber, made by chemically etching a bigger wire, by blasting a larger wire with high energy particles, or by pushing a nanoprobe's tip into the soft surface of a partly melted metal and pulling it back to get a nanowire—like the gooey mozzarella cheese that sticks to your fork in a long thread after pulling it back from a hot pizza.

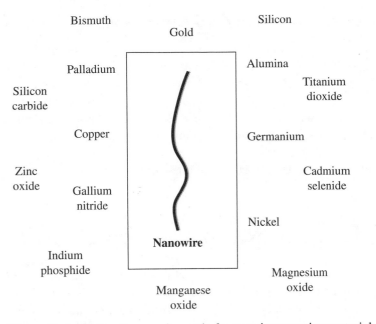

Figure 9-4   Nanowires can be made from various starting materials.

Deposited Nanowires

A deposited nanowire is coated onto a different surface (e.g., a single row of metallic atoms onto a non-conducting surface). This is commonly done with a *vapor-liquid-solid* method using either laser ablated (erased) particles or a gas as source material.

First, the source is exposed to a catalyst (e.g., liquid gold nanoclusters). Next, the source seeps into the gold nanoclusters and saturates them. Once they become supersaturated, the source solidifies and grows outward from the nanocluster. The nanowire's length is controlled when the source is turned off. It's also possible to make compound nanowires with super lattices of alternating materials switching sources during the growth phase. This process produces crystalline nanowires for semiconductor components.

To create electronic components using nanowires, single nanowires are chemically doped to make *p*-type and *n*-type semiconductors. Then, a simple electronic *p-n* junction is made either by physically crossing a *p*-type wire over an *n*-type wire or by chemically doping a single wire with different *dopants* along the length. This creates a *p-n* junction with only one wire. After *p-n* junctions are built, logic gates can be created by connecting several *p-n* junctions together.

Semiconductor and electronic nanowire uses will be important to future computer fabrication since they take advantage of nanometer physics.

Nanowire Conductivity and Applications

Because of their super small size, nanowires have unique electrical properties. Unlike carbon nanotubes, whose electrons travel freely between electrodes, nanowire conductivity is strongly influenced by *edge effects*. These effects come from atoms that are on the nanowire surface and are not completely bonded to adjacent atoms (like those within the nanowire interior). The unbonded atoms can cause defects within the nanowire, which affects its electrical conductivity. The smaller the nanowire, the more surface atoms there are compared to interior nanowire atoms, and the more edge effects come into play.

Quantum Effects in Nanoscale Electronics

Complex electronics and manufacturing of electrical components at the molecular level have led to rapid development and expansion of new nanoscale processing methods. In developing new methods, it's important that nanodimensions are precisely measured and traceable to accepted bulk standards. Kinetic and quantum effects change semiconductor carrier transport and thermal conductivity properties when devices are scaled down ever farther into the nanoscale.

Conductive nanostructures may also be used as quantum standards/tools to make nanoscale measurements. Some labs are working on coming up with a quantum standard for current and capacitance using movement of single electrons. They are also studying atomic wires, polarized nanoelectronics, and magnetic nanostructures.

Future applications could encompass everything from quantum computing to single-particle sensor devices, nanoscale frequency standards, and atom/surface interactions. Quantum effects are fairly weak by normal standards, so new measurement technology is especially critical.

Although a lot bigger than the nanoscale, micro-machines produce nano-effects when they construct materials and interact. For example, nano-effects are important in the development of microchannel arrays used in gene sequencing and other biological assays. Normally, fluids moving through a wide conduit are only slightly affected by the surrounding walls. However, in the case of microfluidics/nanofluidics, super thin fluid stream interactions and membrane effects can cause strange and unexpected reactions.

As far as nanoelectronics goes, a push is on to invent optical switches for fiber networks. Light waves (a few hundred nanometers long) carry gigabytes of data, but the flow stops when light is converted to electrons for processing and switching and then changed back. Since electronic routers cost millions of dollars, there is a lot of economic stimulus to develop all-optical switches. Some of these optical switches take the form of ring routers and lens/micro-mirror arrays that can be

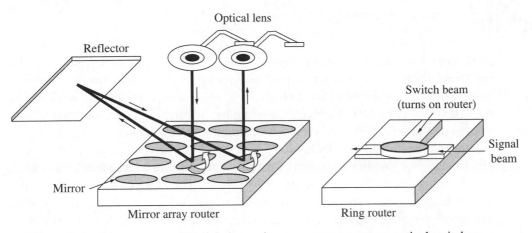

Figure 9-5   Ring routers and lens/micro-mirror arrays can serve as optical switches.

flipped over to divide and direct streams of data-carrying photons. Figure 9-5 illustrates a ring router and a micro-mirror array used for directing information signals.

Bionanosensors

As discussed in Chapter 5, biosensors are important tools that provide selective identification of toxic chemicals at minute levels in industrial products, chemical substances, environmental samples (e.g., air, soil, and water), or biological systems (e.g., bacteria, virus, or tissues) for clinical diagnosis.

By combining the extreme specificity of biological probes and the super sensitivity of laser-based optical detection, biosensors are able to detect and differentiate chemical components of complex systems. Their finely tuned design supplies clear identification and accurate quantification of samples tested.

Added to the already growing arsenal of biosensors are a new group of bionanosensors that use antibody and DNA probes. Antibody-based *fluoroimmunosensors* (FISs), for example, have been developed for carcinogens. Antibodies are placed at the end of a fiber-optics probe or in a sensor (within the FIS) for *in-vivo* and *in-vitro* fluorescence assays. Extreme sensitivity is obtained by using higher laser energies and changing optical signals.

Through a single fiber, the FIS sends increased energy into the sample and collects fluorescence given off from the protein at the end of the probe. Then, when the laser radiation gets to the sensor probe, it excites the sample bound to antibodies on the fiber-optics probe tip. Fast and simple measurements are made with this method, resulting in greater sensitivity of trace chemical and biological components.

Biochips

Biological processes (e.g., cell differentiation, cell division, phagocytosis [engulfing foreign particles] and necrosis [cell death]) are related to movement of cellular components. New methods are being developed to study molecular behavior in important cell processes. Molecular labeling by chemical or DNA methods allows precise tracking of individual molecules using fluorescence microscopy.

The combination of nanotechnology, biology, photonics, and advanced materials allows detection and manipulation of atoms and molecules using nanodevices with a wide variety of medical uses at the cellular level. The fact that the biosensors are nanosized makes measurements in the smallest places (e.g., inside a cell) possible. The development of bionanosensors and *in situ* intracellular measurements of single cells using antibody-based nanoprobes is a big advance.

Using bionanosensors, scientists can test individual molecules and molecular signaling activities in specific cellular locations. Placing a bionanosensor into a cell doesn't seem to bother the cell membrane or alter a cell's normal functioning. This is great news since physicians have wanted to be a "fly on the wall" of cellular processes for years. The ability to monitor *in vivo* processes within living cells would be a huge improvement to their understanding of cellular functions. Just this nanotechnology-enabled improvement alone would transform cell biology.

Biosensors using DNA probes (*biochips*) are a hot new field. As discussed in Chapter 5, protein recognition is brought about through the fine-tuned linking of a nucleic acid strand with a complementary protein sequence. DNA biosensors are useful in areas where nucleic acid identification is involved. These types of sensors could be used to diagnose genetic susceptibility to inherited diseases like hemophilia.

Glucose sensors are probably the most well known biosensors on the market today, since thousands of people with diabetes must be able to monitor their glucose levels throughout the day. Glucose can be tested by using the enzyme glucose oxidase, which combines glucose and oxygen to form gluconic acid and hydrogen peroxide. The sensor detects the amount of hydrogen peroxide formed and current changes that are measured by an electrode.

Bionanosensors allow researchers to make use of and test for biomolecules. They are reaching a point where bionanosensors and nanomedicine will augment biology. Clinicians want to create nanomachines that can carry out and analyze cellular processes normally done by biomolecules or entire groups of cells. Some people call this *cellular engineering*; others call it healthcare of the future.

Electronics innovations using nanotechnology that make smaller, faster, more accurate, and cheaper products are just around the corner (and some are already here). Every industry in our highly computerized society will reap the benefits offered by this amazing field.

Quiz

1. What regulates current/voltage flow and acts as a switch or gate for electronic signals?

    (a) sluice

    (b) transistor

    (c) stopcock

    (d) meter box

2. Silicon, gallium-arsenide, and germanium are used to make

    (a) fertilizers

    (b) plastics

    (c) nasal inhalants

    (d) semiconductor materials

3. What area deals with the behavior of matter at the atomic, nuclear, and particulate levels?

    (a) quantum mechanics

    (b) nuclear fission

    (c) geophysical tectonics

    (d) Edwardian physics

4. The process of imprinting patterns on semiconductor materials for use in integrated circuits is called

    (a) chemotherapy

    (b) photosynthesis

    (c) transduction

    (d) lithography

5. Nanowire conductivity is strongly influenced by

    (a) edge effects

    (b) market results

    (c) solar flares

    (d) chemical composition

6. *N*-type semiconductors with added impurities or dopants cause

   (a) an abundance of quarks

   (b) an excess of electrons

   (c) all charges to be neutralized

   (d) a deficiency of electrons

7. Using extreme-ultraviolet light to carve transistors in silicon wafers will produce microprocessors up to how many times faster than today's most powerful chips?

   (a) 10 times

   (b) 25 times

   (c) 50 times

   (d) 100 times

8. As the computer industry doubled the density of components every year, it became known as

   (a) Kulinowski's Theory

   (b) Paul's Suggestion

   (c) Moore's Law

   (d) Bardeen's Hypothesis

9. When impurities are added to semiconductor materials, the process is called

   (a) scoping

   (b) dubbing

   (c) doping

   (d) masking

10. Antibody-based fluoroimmunosensors have been developed to test for

   (a) mosquitoes

   (b) carcinogens

   (c) low glucose levels

   (d) nuclear waste

CHAPTER 10

Communications

Because of the huge upsurge in communications technology, people call the last 40 years the "Information Age." Keeping in touch with family, friends, and business associates has become incredibly easy. The development of the Internet with instant access to far-flung places has made information exchange something that people only dreamed about earlier.

Well, hang on to your mouse! Information transmission and accessibility are ready to take another leap forward through nanomaterial properties and potential devices. Computer components, as you learned in Chapter 9, can be created at the nanoscale from the bottom up as well as the top down. These methods are also being used in the development of communications components such as cell phones and radar systems.

In fact, many nanoparticle properties important to computer and sensor manufacturing are also related to communications. Some of these (e.g., quantum dots) may be used in both areas. Signal transmission is particularly important in communications and will expand as material properties are better understood.

> *Quantum entanglement* takes place when photons can't be described individually, but instead are qualified as a wave movement, where measurements in one system influence others simultaneously.

Researchers are actively studying the way a single molecule or group of molecules does computations using mechanical, magnetic, or electronic variances. Creating a functional device inside a molecule is a big challenge to molecular electronics researchers, where function is tied to a molecule's physical ability to integrate arithmetic and/or logic.

Molecular electronics cover conceptual, experimental, and modeling challenges. Those challenges bring together molecular architecture, interactions, communication within and between molecules, chemistry, and nanomanufacturing and packaging techniques. In molecular architecture, the biggest challenge is in determining whether a molecule will interact, serve as a switch/transistor, or perform a more complex function (e.g., logic gate or arithmetic and/or logic unit). If an electronic device can be reduced in scale to one molecule per transistor, then molecular technology may come close to semiconductor nanoelectronics.

Researchers have learned that enough quantum resources exist in a single molecule to integrate more than one device function. To make use of this information, many different ways to pick the right molecular models and architecture are being considered.

Quantum Communications

As briefly described in Chapter 9, communications research is currently focused on the area of quantum *entanglement*. When two entangled photons can't be easily separated, their movement is considered to be directly related. In fact, it has been observed that two entangled lower energy photons can be created from a single photon of higher energy.

The potential to transfer quantum information between remote locations (i.e., *quantum teleportation*) is key to the emerging field of quantum communications. Impressive refinements in experimental methods and equipment make it possible for researchers to perform quantum experiments with broad scientific, business, and manufacturing results.

Since quantum entanglement couldn't be observed, analyzed, or utilized until recently, its potential is just now picking up speed. In fact, entanglement offers a great basis for secure (quantum) cryptography, computational algorithms, and quantum teleportation.

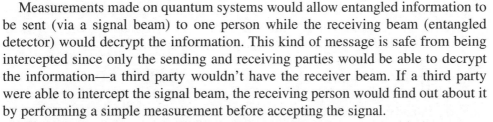

Measurements made on quantum systems would allow entangled information to be sent (via a signal beam) to one person while the receiving beam (entangled detector) would decrypt the information. This kind of message is safe from being intercepted since only the sending and receiving parties would be able to decrypt the information—a third party wouldn't have the receiver beam. If a third party were able to intercept the signal beam, the receiving person would find out about it by performing a simple measurement before accepting the signal.

This seemingly faster-than-light type of communication, along with the extreme miniaturization in computer technology, is reaching the point where researchers will be commonly using quantum physics to describe basic computations. In the nanoscale world, even the theory that describes how computers operate must be revised.

QUANTUM SPIN

Quantum computers will be able to process and store huge amounts of information, a lot more than electronics handle today, because of their ability to work in parallel with all possible answers represented. Traditional computers don't have this luxury and can take a very long time to run through all the operations of a large simulation, encryption, or communications network. Quantum processing will be able to operate millions, even billions, of times faster than today's supercomputers.

Since researchers in quantum computing/communications must use quantum physics to describe quantum mechanical interactions, they are forging new territory. In fact, quantum communications make possible such tasks as breaking previously unbreakable codes, generating true random numbers, and sending messages that tip off the sender/receiver to the presence of a third party.

Quantum computing uses nanoparticle quantum properties to perform computations. One way it does this is with the quantum property of electrons called *spin*. Spin is a fairly complicated concept, but is measured at a value of $-1/2$ or $+1/2$. Researchers think of the $-1/2$ and $+1/2$ spin values in computer terms (e.g. binary language of 1s and 0s, where $+1/2$ is equal to 1 and $-1/2$ is equal to 0). If an electron is like the smallest unit of digital information (bit), then quantum computers manipulating electron spin might have a basic unit called a *qubit*.

The ability to store or transmit information on an electron is quite complicated, but the spin of an electron is not determined until it is measured (i.e., the spin shows properties of both $+1/2$ (1) and $-1/2$ (0).

As mentioned in Chapter 8, electron spin is also affected by light. The wavelengths and type of light (e.g., polarized) used along with the rate of light pulses serve as a program for the quantum computer's electrons. The amazing twist is that since the electron spin is not revealed until you measure it, one computer command might have two operations running in parallel for the 0 and 1. Qubits can also be linked together so that the condition of one affects the others like a domino effect.

In a simple example, 2 qubits can have four possible configurations: (00), (01), (10), and (11). Mathematical operations with qubits could be used to operate on all four different states at once. This process takes us back to the new field of quantum entanglement.

Qubits and spin manipulation will not only speed up computers and communications, but they will make impossibly time-consuming (e.g., billions of years) operations quick and easy. This would allow for fantastic advances in the digital security and database search areas.

It's important to keep in mind that entangled electrons are touchy things. And at the nanoscale, it doesn't take a lot to knock an electron off its path. Any sort of interaction with the environment (e.g., stray photon or vibration) can do it. When this happens, it is known as *decoherence*. When this takes place in quantum computing, an error is created.

A new area, called *quantum error correction*, has begun with researchers looking for ways to protect quantum systems. Several quantum error correction approaches have been suggested by researchers at Bell Labs, Oxford University, University of Toronto, Los Alamos National Labs, and Princeton University. The problem is that for a quantum system to last a long time before decoherence takes place, it needs to be impervious to its environment. The trick is to design a quantum computer that is able to handle strong internal interactions, but that is not impacted by its surroundings.

A lot more research needs to be done to stabilize qubits. Currently, decoherence happens after about 1000 operations. When used to transfer data, this is clearly not an acceptable error rate, but the use of other materials (e.g., nanodots), as described in Chapter 8, may solve this problem.

Chemical Challenges

For communications using data, electrical charges, and energy exchanges (between molecules and the larger world), the big question is how to make fabrication precision and surface positioning (single atom, molecule, or quantum wire) better than 0.1 nm.

To fine-tune and standardize manufacturing specifications that begin in the nanoscale and eventually link with the larger (bulk) world, some step-wise methods have to be figured out. To work toward this goal, very accurate movements are needed. In fact, *atomic precision positioning* is a new research field involved in the transfer of a single molecule from metallic to semiconducting surfaces and then to a much larger external connection. Many research fields are working together to solve various related questions with this physics/chemistry/mechanical/materials/computer engineering problem.

Building a molecular "circuit" board to perform electronic functions or act like a digital logic gate is not the only tricky part of nanocommunications. A big chemistry hurdle for researchers is finding ways to equip a circuit board with side chemical groups that don't add directly to operating functions. At the same time they have to protect other interactions, while assembling/stabilizing specific molecules on the substrate.

For example, the ability to move a molecular board away from the reactive surface while performing scanning tunneling microscope (STM) single molecule manipulation is a problem. To keep a strong connection, side chemical groups must have a strong electrical connection between molecular devices or logic/local leads. Along with adding simple end groups like *thiols* (sulfur-containing chemicals), just adding side groups at all becomes complicated during self-assembly). Because this not an super easy process, new methods will have to be developed.

Fabrication and packaging performed with nanoscale precision is still in its infancy, though various research efforts are proving that it can be done. Eventually, molecules will be able to self-assemble structures such as interconnect junctions. Then, once that snag is smoothed out, single molecular devices or logic gates will need to be moved from the laboratory to the marketplace.

The "Nanotechnology Age" we are entering gives us an opportunity to explore new material properties on the nanometer scale and use them for the greater good. Related tools include methods for controlling atomic and electronic structures on the nanoscale; availability of instruments able to characterize nanoscale materials, structures, and properties across wide length/time scales; and computer power for analysis across the atomic to bulk scales.

So far, the greatest attempts to use synthesis, characterization and modeling tools for the understanding and control of nanomaterial structures have been those with semiconductors and related materials. However, tons of material property combinations will come about giving us many more nano materials.

Size

Size changes often produce new behaviors. For example, if a nanoparticle gets smaller than the normal length for scattering electrons or *phonons* (vibrational energy), it can create new modes of electrical current and/or heat transport. This was seen in the early 1980s in non-superconducting currents of metallic rings and more recently in the super current transport of carbon nanotubes. New properties are also seen when light wavelengths are affected by quantum dot dimensions.

Thermodynamic properties including magnetism, electricity, and superconductivity also change when structures are in the nanoscale range and have either a small number of nanoparticles or a system size equal to the particle size. Systems with particle sizes ranging from a few tenths of a nanometer to about ten nanometers are at the hazy border between quantum and regular ranges.

Currently data storage and information technology progress has sparked increased research in nanoscale magnetism. The magnetism of complex molecules (containing lots of atoms with localized magnetism) has unusual properties. IBM scientists, for example, have been able to improve magnetic resonance imaging (MRI) by directly measuring a weak magnetic signal from a single electron within a solid sample. This result is important in the creation of a microscope that provides 3D images of molecules with atomic resolution. This type of instrument would make it possible for scientists and engineers to study materials (e.g., proteins, pharmaceuticals, integrated circuits, and industrial catalysts) from a much closer vantage point. Knowing the exact spot where a specific atom is located within a nanoelectronic structures, for example, would boost electronics designers' ability to perfect circuit manufacture and performance. Additionally, an electronics/communication approach based on spin, rather than electron charge, in magnetic nanodevices looks like it may be the newest development path.

Changes in nanoscale elements' structural strength, friction properties, and fluid flow (e.g., lubrication) will bring about new design methods for nanodevices. However, there are some drawbacks to working at this scale. As devices are created at the nanoscale, standard properties will need to be studied as everything shrinks toward the nanoscale. Mechanical changes, fracturing, increased surface tension, and greater diffusion and corrosion properties, all related to higher surface-to-volume ratios will be changed. Even heat properties of nanoscale devices will be a far cry from their bulk cousins.

TIME SCALE SHIFTS

It's easy to see how things change the smaller they get, but what about the length of time it takes for events to happen? Scale changes also affect time scales. For example, it takes a lot longer to eat a five course meal than five jelly beans.

In scientific terms, fewer frequencies are needed to move things across shorter distances at a fixed velocity (e.g., photons, electrons, and so on); the shorter the distance traveled, the fewer frequencies are needed to move the object. However, other changes come into play like the increased importance of surface effects compared with internal interactions.

Just as instrumentation such as atomic force microscopy (AFM) has been driven by researchers' need to study the unique properties of nanomaterials, the speed of quantum operations will also require new measuring instruments.

Nano-optics

Nano-optics is the field of optical occurrences in the nano world, where the size of the particle is much smaller than a wavelength of light. While electromagnetic wave effects aren't as big in tight places (less than or equal to half a wavelength), generally, researchers see that nanoparticles oscillate in electric fields at optical frequencies and act like electromagnetic waves.

> The main carriers of energy and coherence in nano-optics are called *surface plasmons.*

These local electrical fields cause hugely enhanced optical phenomena. *Surface enhanced Raman scattering* (SERS) is the most studied and best known of these. When light is scattered from an atom or molecule, most photons are elastically bounced around (*Raman scattering*). These scattered photons have the same energy (frequency) and wavelength as the surrounding photons.

> *Raman scattering* takes place when molecules absorb photons and then the excited electrons scatter energy by giving off both *phonons* (a vibration of the material's crystalline structure) and photons of lower energy than those absorbed.

However, a small fraction of light (around 1 in 100 photons) is scattered at optical frequencies lower than the frequency of the surrounding photons. In a gas, Raman scattering takes place with a molecular change in vibrational or rotational energy.

SERS is normally performed using a silver or gold substrate. Silver and gold are easily excited by a laser, and the resulting electrical fields cause other nearby molecules to become Raman active. When this happens, researchers are able to study chemical bond vibrations within molecules.

But let's back up a bit and figure out what this all means.

QUASIPARTICLES

A quasiparticle describes a single particle moving through a system, surrounded by a cloud of other particles that are being pushed or dragged along by its motion (like the wake of a ship), so that the whole thing moves like a free particle. The quasiparticle idea is one of the most important new ideas in physics because it applies to a wide range of systems.

A quasiparticle with a low excited state and a bare minimum energy level is known as an *elementary excitation.*

In quantum mechanics, an *excited* state of a system (e.g., atom, molecule, or nucleus) is any quantum state with a higher energy than the *ground* state (bare minimum). For example, a hydrogen atom's ground state is a single electron in the lowest possible *orbit* (energy level). When the atom gets additional energy (i.e., absorbs a photon), the electron jumps to an excited state. If the photon has enough energy, the electron can even get kicked out of the atom and the atom is said to be *ionized*.

It turns out that quasiparticle interactions are fairly unimportant at low enough temperatures, but low temperatures allow their flow and heat capacities to be more easily studied. In fact, most many-particle systems are made up of two types of elementary excitations: quasiparticles whose motions are changed by interactions with the other particles in the system, and a combined motion of the whole system. These excitations are called *collective modes* and include things like *plasmons.*

A *plasmon* is a quasiparticle that comes from subatomic particle oscillations that have vertical and horizontal movements.

Plasmons play a big part in the optical properties of metals. In most metals, the light frequency is in the ultraviolet wavelength range, making them shiny in the visible range. Gold and copper, for example, have light frequencies in the visual range, which allows us to see the colors we recognize for these metals.

The importance of plasmons was discovered accidentally (like many scientific breakthroughs) in 1989. A researcher named Thomas Ebbesen at the NEC Research Institute in Princeton, New Jersey, sent light through a sheet of gold foil that contained about 100 million holes (each about 300 nm across). The holes were all smaller than the shortest wavelength (around 400 nm) of visible light. This was like trying to shove a softball through a hole the size of a ping-pong ball. In fact, optical quantum theory predicts that only around 1/1000 of the light shined on the holes should be able to get through. But when Ebbesen shined the light on the foil, *more than 100 percent* of the light got through! More light actually came out the other side than went in! He performed the experiment several times and checked different aspects of his findings, but he reached the same results. It was so weird that he thought it was a fluke and didn't even report his findings.

In 1998, however, Peter Wolff, a theoretical physicist, joined NEC and heard about Ebbesen's nanoscale holes/light experiments. Since Wolff already knew how electrons acted on metal surfaces (they ripple across on surfaces in waves), he tried the experiment. He found that the electrons behaved like waves, not particles, and this special nanoscale effect let more light (at the right wavelengths) through the holes than would be expected in the larger world.

Now Plasmons are being considered for use in transmitting computer information, since they send information a lot faster than electrons, but can be channeled along conducting paths like standard computer chips.

Nanolens

A team of researchers from Georgia Tech and Tel Aviv are working on a way to focus light to a point only a few atoms in diameter and amplified a million times. By using a series of several metal nanospheres with decreasing sizes and separations, light can be focused to an extremely intense point. It's a lot like using a lens to focus sunlight and burn a spot on a piece of paper. This nanosphere method could be used in nano-optical detection, nanomanipulation of single molecules/nanoparticles, and other applications.

A simple model of a *nanolens* process uses a row of three successively smaller nanospheres from 50 to 5 nm in diameter. Figure 10-1 illustrates how energy is focused down with smaller and smaller lenses. When the largest sphere is illuminated, a wave of vibrating electrons (plasmons) moves across the nanosphere's surface and creates an oscillating electric field (a lot like Ebbesen's and Wolff's earlier work). This field moves down through the range of spheres to a point between the smallest and next to smallest spheres. When most of the light energy reaches this point, it has gone through a million-fold intensity increase.

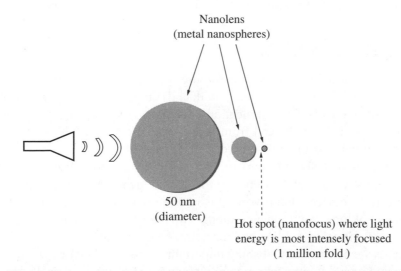

Nanolens
(metal nanospheres)

50 nm
(diameter)

Hot spot (nanofocus) where light
energy is most intensely focused
(1 million fold)

Figure 10-1   a nanolens produces an electric field that progresses through smaller and smaller spheres to achieve a million-fold intensity increase.

This amplification process has benefits for various kinds of spectroscopy. Surface plasmons might be focused in a laser type device called a *spaser* (surface plasmon amplification by stimulated emission of radiation). A spaser works with a metallic nanoparticle and active medium that has semiconductor quantum dots.

Remember when everyone got excited about fiber-optics? Nano-optics may offer a leap that communications hasn't seen since fiber-optic cables hit the scene. With this type of light energy research beginning, science and engineering's ability to advance communications and information transmission, as well as develop fancy encryption methods, have becoming a reality.

INTERACTIVE MODELING AND STANDARDS

As described in Chapter 4, modeling and large-scale computer simulation is important in understanding nanoscale properties and effects. Since the relationships among electronic, optical, mechanical, and magnetic properties of nanostructures and their size, shape, topology, and composition are not really well understood (except for the simplest semiconductors and carbon nanotubes), a lot more modeling must be done.

In nanoscale systems, heat and quantum fluctuations are about equal to devices' *activation energy* (energy needed to overcome inertia). Statistical models have to include these changes in order to describe the big picture accurately. Computer simulation models using quantum methods are needed to check out nanoscale device performance. Computer simulations will also play a huge role in describing materials at the nanoscale as well as in designing new nano materials and products.

SIGNAL TRANSMISSION

While technological innovations are forcing big growth in the communications industry (e.g., DVD, wireless, and broadband), they are also underscoring the importance of standards. Communications standards are essential for sending messages through various media and from one device to another. Unless standardization exists, communication will not work well between systems.

Communication is about being "on the same wavelength." When a misunderstanding occurs, we sometimes say that the people involved are *out-of-sync*. When communications devices don't work, this is often the problem, literally. Precision measurement of time and frequency, along with many other measurement tools and standards, are critical to sending and receiving communications.

Standards agencies are establishing important markers in the forward progress of nanotechnology and its impact on communications. Frequency signals needed for radio, television, telephone, and Internet communications, as well as navigation and space exploration, are interrelated. Optical fiber and wavelength calibration standards for optics and telecommunications are going to have to keep with new nanotechnology-related devices in order to make everything work well together.

Networks

Global communications are ever dependent on computer networks for electronic commerce, health care, education, science, and entertainment. Nanotechnology-based networks will require network protocol designers, engineers, programmers, and testers to review systems/components with new test methods and measurement technologies. In high-speed networks, Internet technologies, multimedia networking, and wireless networks, new standards must be developed to meet the need of upcoming quantum scales and speed.

Wireless

Wireless technology has become the new "wow" technology for much of the general public. As communications have moved from the early pagers to cell phones and personal organizers, keeping in touch has become much more important in many societies. Some might argue whether or not this is a good thing, but it is reality!

The tiny, intricate components that make up cellular telephones, pagers, and other miniature communications devices have changed our lives and relationships in major ways. Consequently, we are at the point now of being annoyed when someone's personal ring tone interrupts everything from movies and school concerts to board meetings and church services.

For consistent delivery of clear communications, key parts have to perform in a wide range of temperatures. Nanotechnology has improved the wireless industry through the development of ceramic materials (e.g., wireless device filters, resonators, and oscillators) as well as ceramic thin films. In fact, in 2003 the Larta Institute, an independent, private, nonprofit corporation in California, jointly awarded "The Nano Republic Award for Most Promising Application" to Hybrid Plastics and Shea Technology Group, Inc., for the first-ever *Nanostructured hybrid*, a nanoscale inorganic-organic chemicals technology. Shea Technology Group is a long-time player in the wireless technology market.

Computer Security and Public Safety

Many people in small town America don't bother to lock their doors at home or worry about car-jacking when traveling. If only the rest of us were so lucky. With urbanization and high-speed lives comes everything from run-of-the-mill e-mail scams to identity theft and fraud. Many of these crimes involve computers and communications.

Investigation and forensics agencies need high-powered reference tools for investigating and prosecuting software piracy, copyright infringement, child pornography, and related crimes involving information files. This includes systems that classify and match fingerprints, as well as facial feature recognition technology used by law enforcement agencies.

For police, fire, and emergency medical personnel to do their jobs (e.g., handling fender benders to responding to major emergencies like energy blackouts or hurricanes), officials need high-speed, reliable communications systems that help them communicate quickly. Important communications systems need encryption and secure processing to avoid being a target for high-technology information crimes. Optical nanotechnology devices may soon meet this need with advances in quantum communications.

Video Advances

It has been said that a picture is worth a thousand words. If that's true, then millions of digital bits that are replacing film reels and video cassettes must be worth volumes and volumes. Along with progress in communications and information distribution, aided by quantum entanglement and nanotechnology, companies are producing higher resolution and truer colors on tiny handheld computers as well as huge stadium-size displays.

One of the big hits of the holiday gift-giving season a few years back was the introduction of the flat-panel display. Nano-optical arrays will keep improvements in entertainment and other visual areas coming for a long time. Additionally, developing nanoscale colorimeters and nanoparticle tunability will greatly improve colors in photonic imaging displays.

Information Storage

As with the electronics industry, the data storage field is moving toward the nanoscale. By shrinking components to below 1/10,000 the width of a human hair, developers are able to make faster computer chips with more capacity. However, the technology for making nanoscale devices is still being developed, and the smaller the components get, the more expensive they are to produce.

Most manufacturers believe that the convenience of electronic devices depends on how much information can be crammed into super small amounts of space. Major advances in digital data storage have been suggested using magnetic data storage (e.g., high-density and high-speed magnetic recording, magnetoresistive sensors, and memory elements).

Researchers are looking into new storage methods using magnetic nanomaterials. They are also trying to figure out how the physical and magnetic structure of magnetic recording material influences storage capacity.

In 2004, IBM announced a major advance in data storage with its Millipede system, which uses a 32×32 array of 1024 atomic force microscope (AFM) probe tips to make 50 nm indentations in a polymer. (Look back to Chapter 4 for how an AFM works.) The Millipede micromechanical device uses *cantilevers* to read and write to nanoscale pits (indentations). Figure 10-2 illustrates a cantilever being deflected by a change in current. The AFM tip interacts with the pits by thermal coupling and

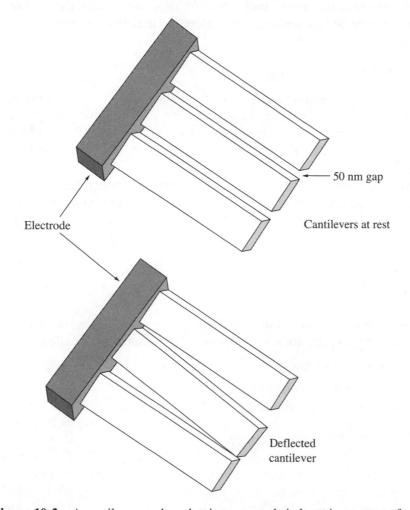

Figure 10-2   A cantilever reads and writes nanoscale indentations on a surface.

either melts a nano pit into the polymer surface or deletes the nano pits by heating the entire recording surface. This creates storage/retrieval systems that could hold one terabit per square inch—40 times the data density of current technology. And it would use less energy than most magnetic methods.

At National Institute of Standards and Technology (NIST), a *pulsed inductive microwave magnetometer* (PIMM) is being developed to monitor picosecond (10^{-12}) measurements of super fast magnetization switching. This instrument will help researchers test highly magnetic nanomaterials for recording data in the extremely small bits (≤ 160 nm² per bit) needed for high-speed recording heads. Using PIMM, materials researchers can check out the composition and growth conditions that bring about high-speed response, while developing magnetic memory that can read and write data at speeds over 1 billion bps. At these future rates, the contents of the entire *Encyclopedia Britannica* could be stored in less than a minute!

QUANTUM POTENTIAL

Quantum computing and communications will allow tremendous processing power through the ability to work in multiple states and perform tasks and all manner of analyses at the same time. Research leaders in this area are IBM, Massachusetts Institute of Technology, Oxford University, and Los Alamos National Laboratory, to name a few.

Quantum computing and communications are expanding the cornerstones of information processing through quantum physics, the most applicable tool of predicting molecular interactions currently known.

Quiz

1. When photons can't be described individually, but as a wave movement where measurements in one system influence others, it's called

   (a) quantum spaghetti

   (b) quantum entanglement

   (c) phototropism

   (d) solar flares

2. When nanoparticles get smaller and smaller, it can lead to new modes of

    (a) gaming devices

    (b) collaborative research

    (c) electrical current and/or heat transport

    (d) steam locomotion

3. In a nanolens, when the largest sphere is illuminated, a wave of vibrating electrons (plasmon) results in an

    (a) accordion sound effect

    (b) amplified heat wave

    (c) augmented visual acuity

    (d) oscillating electric field

4. To avoid being a target for high-technology information crimes, communications systems need

    (a) reliable cooling systems

    (b) copper-base wiring

    (c) encryption and secure processing

    (d) iris scan identification systems

5. A quasiparticle resulting from plasma oscillations, with vertical and horizontal components, is called a

    (a) nebuton

    (b) xenon

    (c) plasmon

    (d) plasmid

6. What does the Millipede use to read and write to nanoscale pits?

    (a) cantilevers

    (b) osmotic pressure

    (c) nanolens

    (d) freon

7. The basic unit of a quantum computer manipulating electron spin is a

(a) que

(b) qubit

(c) microbit

(d) cubit

8. The Millipede could help create storage/retrieval systems that could hold how many times the data density of current technology?

(a) 10

(b) 20

(c) 30

(d) 40

9. Elementary excitations that are carriers of energy and coherence in nano-optics are called

(a) surface plasmons

(b) subterranean spasmons

(c) quarks

(d) transistors

10. Quantum computing and communications have the potential to provide tremendous processing power through their ability to

(a) become very small and compartmentalized

(b) focus research funding

(c) harness fossil fuels

(d) work in multiple states

CHAPTER 11

Energy

Imagine what the world would be like in 2050 without renewable energy. Nearly all the extractable oil on earth will have been burned to fuel automobiles. We will have squandered almost all of the earth's natural gas on generating electricity and condensing it into liquid fuel for cars and planes. We will be burning the remaining coal laced with metals and contaminants that can't be easily removed. The atmosphere will have passed the 750 ppm mark for carbon dioxide (CO_2) concentration. The oceans' shorelines will have moved inland about 50 miles, drowning most of the world's coastal cities and creating many new ones that will be much more crowded. The electricity grid will be old and totally unreliable, and most people will have electrical power only a few hours a week. Billions of people will have migrated on foot to escape sweltering heat (in former temperate zones) made unbearable by the lack of reliable power for air conditioning.

Poverty and disease will be everywhere, determined by the appalling lack of widely available, clean, cheap energy. Energy will be a luxury available only to the rich, who spend much of their wealth trying to protect themselves from the hoards of energy hungry poor.

We can't imagine the world's leaders allowing such a scenario to occur. But in 2006, we seem to be on a path of wishing for new technical solutions, rather than finding money for research and the development of energy alternatives, which must be done before fossil fuels become too expensive and scarce to support the world's needs. This chapter presents a path to the solutions needed by 2050.

Nanotechnology discoveries are causing a domino effect of innovation across nearly every science and engineering field. As more and more technologists learn the fundamentals of nanotechnology, and more unusual nanoscale properties are understood, more powerful uses are being imagined. Perhaps the most globally exciting nano application is in the area of energy. Humanity's future prosperity and energy availability, as well as the quality of the global environment, is the most important area that will be affected by nano applications.

The late Dr. Richard Smalley, a champion of technology and alternative energy options for many years, was convinced that new nanomaterials were critical to having enough energy to satisfy future world needs. He saw nanotechnology as the way to distribute energy around the world to where it is needed. An advocate for aggressively addressing tough energy demands, Smalley believed that people needed to understand and apply nanotechnology to every energy issue.

In order for everyone in the world to benefit from life-enriching new technologies, energy needs must be met. With fossil fuel supplies dwindling, we must find economical alternatives soon or we will all find ourselves going fast.

Energy

Before we can see how nanotechnology can be used to solve our energy problems, we need to understand what those problems really include. So, let's launch into the energy story.

What are the most important problems facing humanity for the next 50 years? The list is long but linked by a common denominator: energy. Table 11-1 shows the top ten problems facing humanity today.

Usually, people think energy comes in around fourth or fifth. But if you think about it, what would happen if quantities of inexpensive, environmentally friendly, and widely available energy were in abundant supply? It would solve a lot of humanity's material problems!

Look at a another big problem—availability of clean water. There's a lot of water on this planet (over 70 percent of the Earth's surface, in fact), but it's salty and not always accessible. What does it take to fix the water problem? It takes energy, and a lot of countries can't afford to clean up or pump the water they need. How about food? There's a lot of arable land (land fit for cultivation) in the world, but we don't have water to irrigate crops and/or energy to provide clean water everywhere.

Table 11-1 Most of the world's top problems could be solved with more available energy.

Ranking	Problem
1	Energy
2	Clean Water
3	Food
4	Environment
5	Population
6	Disease
7	War/Terrrorism
8	Poverty
9	Education
10	Land

What about the environment? Clearly a lot of our environmental problems result from the kind of energy we use, now mostly fossil fuels, like oil, natural gas, and coal, but also wood and animal waste for heating or cooking. These fuels produce a lot of CO_2, soot, and other atmospheric contaminants that dirty our air and cause the temperatures around the world to increase (global warming).

Poverty, almost by definition, is a lack of energy. Of the 6.5 billion people on the Earth, 2 billion have no access to electricity, and another 2 billion use few fossil fuels, only biomass (e.g., wood and animal waste). The rest of the Earth's people use all the rest of the energy. (Americans, for example, make up 5 percent of the population but use 25 percent of the energy.) The people without access to energy are having problems with the rest of us using so much, which incites unrest and political tension. The great economic differences between the rich and the poor contribute to many of the disputes among people and nations.

How about terrorism and war? Have we ever fought a war over energy? The Gulf War and burning oil fields come to mind.

Consider disease. A lot of disease comes from contaminated water, so if we have the energy to provide/clean up water sources (already on the "fix it" list), then disease levels will drop in a big way. Education is a bit of a stretch, but it's tough to learn on an empty stomach. Ask any student.

The number of humans on Earth is expected to expand to 8 to 10 billion by 2050. Democracy, freedom, and population might be less obvious problems than energy, but by solving the world's energy problems, nations would be in a better position to master the difficulties facing humanity.

Which of these factors, besides energy, can solve any or all of these problems? Thinking through each factor carefully will tell you that only massive population reductions can make any real difference. In fact, the worse wars and most fearsome plagues eliminated only a small fraction of the world's population. We can't evict 5 billion people from the Earth so that the remaining people can live on the resources that are left. If we moved population to the top priority on the list, and allowed only 1 billion people to inhabit the planet, would we still have all these problems? Probably not; most of the problems would disappear or decrease drastically. But what would we do with the extra 5.5 billion people? Eviction notices are kind of hard to come by for 5.5 billion people.

If you've been to the bookstore lately, you've probably noticed a lot of books on energy, such as *Out of Gas*, *The Oil Factor*, *The End of Oil*, *The Final Energy Crisis*, and *Beyond Oil: The View from Hubbard's Peak,* to name a few. Most discuss whether or not world oil production has already peaked. Nearly all agree that with world consumption rising, oil production will peak by 2010 if it hasn't done so already.

M. King Hubbard, a geophysicist at Shell Oil, calculated oil production levels based on total extractable amounts. This 1957 analysis showed the current and future extraction of oil over time. The peak (later known as Hubbard's peak) showed when half of the world's extractable oil has been used. Hubbard predicted U.S. oil production to peak in 1970. It did. Figure 11-1 illustrates the United States production peak Hubbard predicted.

Twilight in the Desert talks about the Saudi Arabian oilfields. It's the first external analysis showing how the world's largest oil fields have probably already peaked, so soaring oil and gasoline prices are probably here to stay as demand grows faster than oil-rich countries can pump it out of the ground.

We're also facing a problem with atmospheric CO_2 concentrations. It's pretty well acknowledged by 99 percent of the scientists today that current CO_2 concentrations are going to cause severe problems over the next 100 years. All fossil fuels produce CO_2 when burned, so if we continue to use more and more oil, natural gas, and coal, we are going to have to find a way to capture and hold on to the CO_2. Many companies are looking at ways to sequester CO_2 in deep wells or in the ocean, or to change the CO_2 into minerals such as calcium carbonate. The problem is that we'll have to store many gigatons of CO_2 for more than 100 years to lower the CO_2 level in the atmosphere significantly.

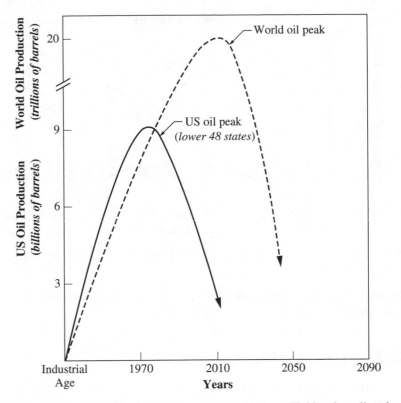

Figure 11-1   U.S. oil production peaked in 1970, as Hubbard predicted.

Availability

Energy is not widely available to many people of the world. About 2 billion people use no electrical power because they have no way of making or getting electricity. Another 2 billion people (some overlap the first 2 billion) use only biomass (i.e., wood or cow dung) for heating and cooking.

Figure 11-2 illustrates the shape we are in as far as oil usage. We're using a lot of oil and a lot of gas, and that's a big problem. We're still burning rail cars of coal and demand is rising. As we approach 2050 and population pressure increases, demand is going to keep going up. In fact, current estimates are probably very conservative, and if we've missed our guess on influencing factors, we're going to have even bigger problems.

For example, China's energy demand isn't growing by the worldwide rate of about 2 percent per year; instead, it has been growing by roughly 20 percent per year. Since China has more than 1 billion people, the average global growth rate for energy is increasing, not staying steady. In fact, it gets worse as more highly

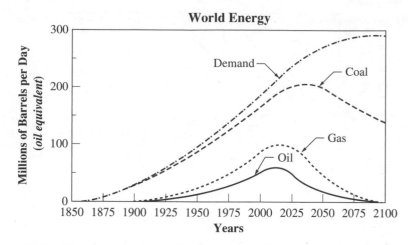

Figure 11-2 World energy demand is rising faster than current fossil fuel sources.

populated countries want cars, trucks, and SUVs! Figure 11-3 shows how much of the world's energy resources come from fossil fuels.

TERAWATT CHALLENGE

The term *terawatt challenge* summarizes the energy problem. Today, the world is using about 210 million barrels of oil equivalent per day. We actually consume around 85 million barrels of oil per day and the rest of it comes from coal, gas, some fission, biomass (cow dung, other animal products, and wood), a little bit of hydroelectric,

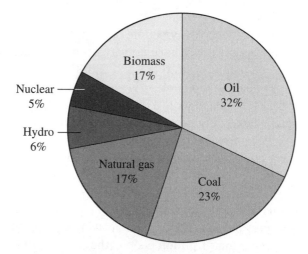

Figure 11-3 More than half the world's energy choices come from fossil fuels.

and almost no renewable resources. So the twentieth century's prosperity was mostly based on oil. What's the energy currency of the twenty-first century?

By 2050, we're going to need at least twice the amount of energy that we are burning or consuming today. Our estimates may be too conservative; in fact, we may need four times the energy. Assuming that twice as much energy is needed, it equates roughly to 10 to 15 terawatts of energy per year.

> A *terawatt* is a million million watts (i.e., 10^{12} watts) and is roughly equivalent to a million barrels of oil equivalent (MBOE).

We don't have that capability today. If you look at availability, oil, coal, and gas will represent a much smaller percentage of that mix. Global demand will require that energy come either from fusion, fission, biomass, hydroelectric, or some other renewable energy. How in the world are we going to make that happen in a few decades? That's the big question!

Efficiency

Another future view comes from Amory Lovins of the Rocky Mountain Institute. Lovins says that we can get the energy we need through efficiency and advanced materials. For example, if we make our cars the same size they are today, but construct them out of advanced composite materials, our energy demand would be less. New cars could be built to save up to 69 percent of the gas consumed today. The same thing could be done with trucks and SUVs (saving 65 percent) by totally redesigning them to use stronger composite materials and making them much more fuel efficient.

Boeing and Airbus are making progress with advanced materials. Their airplanes are designed for strength, efficiency, and economy, not size. For example, the planned Boeing 787 dream-liner is being designed with the future in mind. That's the way aviation needs to go in order to stay alive.

Can we conserve our way out of the problem? Probably not. Many scientists and industry experts don't think conservation will solve the problem. There's not enough renewable power. It'll be pretty tough to get enough biomass for energy and still feed the world properly.

Alternatives

Is hydrogen the answer? The United States government is paying a lot of attention to the hydrogen economy, but hydrogen isn't a primary fuel. Somehow, we have to make a lot more hydrogen before it can be burned.

The University of Nevada, Reno, Materials Nanotechnology Research Group, under the direction of Dr. Manoranjan Misra, Professor of Materials Science in the Department of Chemical and Metallurgical Engineering, has developed titanium dioxide nanotube arrays for generating hydrogen by splitting water using sunlight. Once the process is scaled-up to generate a lots more hydrogen from water, it will have great potential as a clean energy resource.

This new method splits water molecules, creating hydrogen energy more efficiently than currently available. The creation of carbon nanotubes is done by a simple electrochemical method. University scientists add different nanotube materials to increase the water-splitting efficiency and use free sunlight to boot!

Misra says that 1 trillion nanotube-holes can be added in a solid titanium oxide substrate, which is approximately the size of a thumbnail. Each of these holes, a thousand times smaller than a human hair, acts as a nanoelectrode. The hydrogen project also stores hydrogen in nanoporous titanium and carbon nanotube assemblies. These nanomaterials are powerful enough to maintain hydrogen for use in vehicles.

In the May 2004 issue of *Physical Review Letters*, a team from Los Alamos National Laboratory found that quantum dots produce as many as three electrons from one high energy photon of sunlight. When today's photovoltaic solar cells absorb a photon of sunlight, the energy gets converted to one electron, and the rest is lost as heat. This nanotechnology method could boost the efficiency from today's solar cells of 20–30 percent to 65 percent.

NUCLEAR

How about nuclear energy? Fission generates a lot of interest, but to solve the terawatt challenge many nuclear reactors will need to be built. We can't afford to build/operate just regular reactors—breeder reactors are necessary, because there's not enough available uranium to get the terawatts needed. So to get enough power by nuclear alone, how many breeder reactors need to be built to provide 2050's 10 terawatts? The number will surprise you! It's 10,000. To provide 10 terawatts of energy by 2050, one breeder reactor would have to be put into use every day for 27 years, starting now. It could be done, but that's a whole lot of breeder reactors to build. It would be an impossible mission that the world could do it in the next 50 to 100 years.

Nuclear fusion would be a great solution for energy generation. It may not ever get developed, but if it does, it will definitely be used. The cost, however, will be considerable to build fusion plants. There are also a lot of risks and problems related to the storage of radioactive waste from nuclear plants. It's a serious environmental concern.

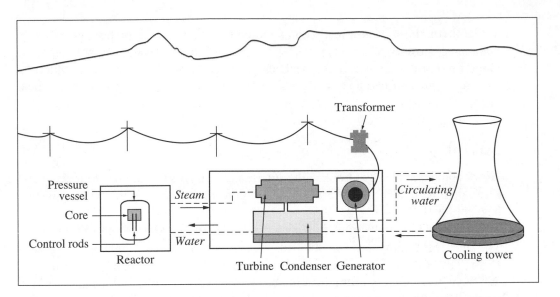

Figure 11-4   Nuclear power transmission can be improved through use of nanomaterials.

Another idea is to use the unique properties of nanomaterials throughout the process of generating/distributing power from nuclear plants. Figure 11-4 shows a plan for a nuclear plant and the transmission areas where nanotechnology-based materials might make the process more efficient.

GEOTHERMAL

Geothermal energy is also a good solution. *Geothermal energy* is created when underground heat is transferred by water to the surface. Think of a wet sauna where rocks are heated and steam is produced when water is poured over them.

To harness geothermal energy, bore holes are drilled into hot underground locations and the super heated groundwater is brought to the surface. But holes must be drilled to retrieve the available water, and then more holes must be drilled to get more. Basically, it takes a whole lot of holes to achieve geothermal on any large scale. And the cost doesn't drop the more drilling you do.

According to the 2005 World Geothermal Congress, 72 countries reported using geothermal energy for direct uses (e.g., space heating, snow melting, aquaculture, and greenhouse use) providing over 16,000 MW of geothermal energy and 24 countries used geothermal to produce 8,900 MW of electricity directly. Reykjavik, Iceland's capital, is heated completely by geothermal energy from volcanic sources within the Mid-Atlantic Ridge that intersects the country. In fact, water reservoirs with temperatures of 80°C–180°C were tapped to provide reliable and low cost heating for homes, businesses and industry.

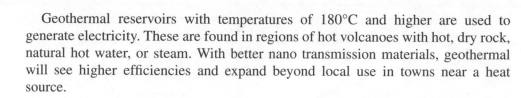

Geothermal reservoirs with temperatures of 180°C and higher are used to generate electricity. These are found in regions of hot volcanoes with hot, dry rock, natural hot water, or steam. With better nano transmission materials, geothermal will see higher efficiencies and expand beyond local use in towns near a heat source.

SOLAR

How about solar energy? Look at the sun (not directly, as it will hurt your eyes) and compare it with the Earth in terms of size. Figure 11-5 compares the sizes of the two heavenly bodies.

How many terawatts of energy come from the sun roasting the earth every day? The answer is a whopping 165,000 terawatts. By comparison, this makes 10 terawatts seem pretty trivial. All we have to do is capture 10 terawatts of energy from the sun and the world's energy problem is solved. We can save our oil, fuel the airlines, and use it in manufacturing.

Is solar practical? Yes and no. The U.S. uses about three terawatts per year in 2006. To satisfy the whole world's energy needs of 20 terawatts, solar cells would

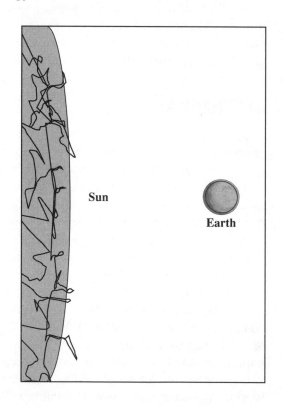

Figure 11-5 The sun produces more energy daily than the Earth could ever use.

need to be built to cover a good part of Texas, Oklahoma, Kansas, Colorado, and New Mexico. But that's not practical.

What's the biggest problem with solar power? Storage. Of course, the sun doesn't shine at night or when it's overcast, so the energy has to be stored in some way so that it can be used when the sun isn't out. And when the sun doesn't shine every where equally, you've got to be able to transmit the electricity to the places where it isn't shining.

Solar energy is the best energy option for the southwestern United States. If we devoted a large area to solar now, we could satisfy the total electricity needs of the United States. But the government is investing less than $100 million a year in solar energy research. For solar to step up to the plate and meet rising energy demands, the government needs to invest in this renewable energy and energy technology.

NATURAL GAS

What about other chemical processes such as natural gas? Unfortunately, this brings us back to the problem of rising CO^2 levels and the cost of distributing the natural gas. In the July 2005 issue of *Scientific American*, Robert H. Socolow describes in "Can We Bury Global Warming?" how William Shakespeare breathed 280 molecules of CO_2 out of every million molecules entering his lungs. Today, our lungs take in 380 molecules of CO_2.

The hot trend is toward *carbon sequestration* (storing carbon underground or in the deep oceans) instead of sending it into the atmosphere via auto exhaust and industrial smokestacks. Energy alternatives would concentrate on using energy more efficiently and substituting renewable energy or nuclear sources for fossil fuels.

Clean coal presents yet another possibility, but carbon sequestration issues and cost are huge barriers unless nanotechnology can overcome these problems.

SMART GRID

To make a "smart energy grid" for the United States and other countries around the world, power cables need improvement. With new super conductive and low-loss electrical cables, we can change the way we create and transport energy. Dr. Smalley described this as the *distributed storage and generation grid* (stor-gen grid). The world's energy system needs to change from a mass-driven system (transporting coal, oil, and gas from place to place) to a transmission system (transmitting electrical energy through cables). Figure 11-6 highlights how energy delivery can be updated.

Nanomaterials (e.g., carbon nanotubes) offer a solid path to updating energy transmission. Since they are strong and conduct electricity 6 times better than copper, their use makes sense as an energy option. Their size is another advantage,

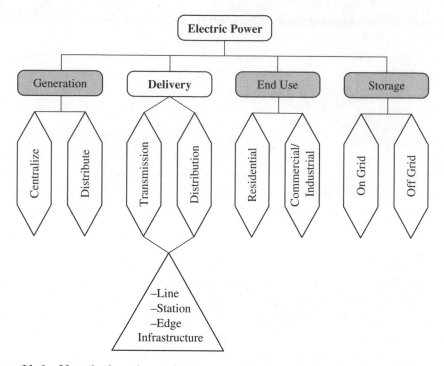

Figure 11-6 If methods and materials were updated, energy transmission will be much more efficient.

especially in large cities that have run out of underground real estate. In New York City, the underground utility corridors are so full of copper wires, it is joked that the city has the greatest supply of copper in the world.

Carbon nanotube cables would be small enough to be added to existing utility corridors. When demand out-paced supply, metropolitan utility companies would have an alternative to buying expensive real estate.

The current United States power grid is old. In fact, most of its main components are 15 to 20 years old or older. The grid needs to become more like the Internet: big, robust, and resistant to terrorists, while still providing local storage (e.g., batteries, flywheels, hydrogen, and so on). During a blackout along the United States' East Coast on August 14, 2003, a lot of people were without electricity for nearly of a week. Transmission and distribution were disrupted without efficient alternatives. Change must begin immediately if blackouts are not to become the norm as other countries have experienced (e.g., Italy, England, Canada, and Scandinavia).

Table 11-2   The twentieth century power grid is much less adaptive than a smart grid.

Twentieth Century Grid	Smart Grid
electromechanical	digital
1-way communications	2-way communications
central generation	distributed generation
radial topography	network
few sensors	integrated monitors and sensors
sightless	self-monitoring
manual restoration	self-healing and restoration
vulnerable to failures and blackouts	adaptive protection and islanding
manual equipment check	remote equipment monitoring
committee and phone emergency response	decision support systems, predictive reliability
low power flow control	all-encompassing control systems
low price information	complete price information
few customer choices	diverse customer choices

> ***Energy blackouts*** are commonly the result of imbalances in electricity supply and demand.

Local generation is also important. No matter how electricity is generated (e.g., solar, nuclear, geothermal, and so on), we need to accommodate all possible processes. A robust electrical grid made from nanomaterials or smart materials can do that. The beauty of a smart grid is that everybody plays in the new game. Table 11-2 compares the twentieth-century grid with a twenty first-century smart grid.

STORAGE

We need to figure out how to store electricity better. If homeowners use solar completely, they will need a home storage device that will store enough power for about a week's worth of use. If or when the system goes down, a self sufficient option would be ready. We can accommodate that now using lithium batteries, but they would take up the square footage of a modest house and cost around $50,000 or $60,000. We need to come up with a storage device that's about the size of a washing machine and that costs about a $1000 if we hope to transform the home energy picture.

How will storage capabilities be accomplished? A smart grid may be the answer. The United States National Energy Technology Laboratory described grid

Table 11-3 A smart grid possesses strengths in a variety of areas.

Grid Modernization Points	Grid Capability	Description
1	Self-healing	Able to detect, analyze, respond, and restore itself rapidly from problems/impacts
2	Incorporates the consumer	Able to include consumer equipment and behavior in grid design and operation
3	Attack resilient	Mitigates and withstands physical and computer attacks
4	Power for 21st century users	Provides quality power needed by industry and consumer needs
5	Variety of generation options	Works with a wide variety of local and regional generation methods (e.g., green power)
6	Enables maturing electricity markets	Allows competitive markets
7	Promotes optimization	Good IT and monitoring to optimize capital assets constantly while minimizing operations and maintenance costs

modernization as needing seven key components. These important factors include not only generation, but energy transmission and delivery factors. Table 11-3 lists these seven components.

Carbon Nanotubes

Engineers are excited about nanotubes because they are the strongest material ever made in the universe by anyone, anywhere. No element on the periodic table can be combined in a better way to make a stronger material. That's a very bold statement, and only a Nobel Prize winner can get away with saying that.

If Dr. Smalley was right, this miracle material with its interesting properties can be used in thousands of different ways. Since it has selectable electrical properties, it could be used to make metallic materials that conduct electricity better than copper. It could be formed into semiconductors and used to build new computers. Nanotubes have the thermal conductivity of diamond, the chemistry of carbon, and the scale and perfection of DNA; they are ultimately the most versatile engineering materials ever imagined.

Researchers can roll a nano sheet of graphite into a nanotube. They can match up the atoms and create a perfect cylinder. We saw that back in Chapter 1. A nanotube conducts electricity better than copper, but it is only a nanometer in diameter and up to 5 inches long. This fiber is around 10–100 times stronger than steel, Kevlar, or any of the strongest fibers ever made.

When you look at about a gram of raw carbon nanotubes from the reactor at Rice University, it looks a bit like black lint. In 2004, that amount cost $1000 dollars. Now, a gram of nanotubes processed by Carbon Nanotechnologies Inc. (a nanotechnology spin-off company in Houston) is $375 a gram. That's about $160,000 a pound, and the price is projected to come down to about $10 a pound.

Researchers want to make the nanotubes into *quantum wire* with conductivity 10 times that of copper, one-sixth the weight, and stronger than steel with zero thermal expansion along its length (it doesn't sag in warm weather). It also doesn't cause power outages.

Dr. Smalley envisioned nanotubes placed end-to-end and parallel to each other that would carry electricity faster and cheaper than today's power lines. Nanotube lines would be much lighter than copper, steel, or aluminum, and electrons could easily jump from one line to another. Superior heat conductance and other nanoscale properties would prevent electricity loss during transmission.

For energy transmission, quantum wires could serve as conductors thousands of miles long and with almost no energy loss. This would revolutionize the energy grid. Making long fibers is the key and researchers have already spun fibers from nanotube materials.

So the initial problem of processing has been solved. Now carbon nanotubes need to be separated into pure, metallic conductors and standardized in large batches. Again, research is making this happen.

Research and Development

The biggest single challenge for the next few decades is creating and distributing energy for 10^{10} people. We need a minimum of 10 terawatts per year from some new clean energy source by 2050. For worldwide prosperity and peace, energy needs to be cheap, widespread, and available. We can't make it using current technology alone.

We still have a long ways to go. Energy research needs more funding. Smalley had an idea for research to get a nickel for every gallon of oil product used in the U.S. That alone would generate $10 billion a year of extra research money for energy. The United States puts a boat load of money into energy production and distribution, as well as import, but not into energy research. Major investment would create a wealth of new technologies and could solve the energy problem by 2020.

The world energy landscape will change dramatically over the next 50 years. Technology will provide options. Leadership of the future world energy economy is up for grabs. The United States is currently focused on oil, gas, and hydrogen, although that may be changing since President George W. Bush signed the 2005 energy bill. The oil and gas energy industry (the largest industry in the world) invests the lowest percentage of any industry in research and development. Considering the size of the problem, that seems fairly short-sighted.

Investment

Where is the intellectual property going to come from? Where are solutions to future energy needs going to come from? Who's going to own the technology? And who's going to be buying it from whom? Those are questions we really ought to be asking ourselves.

Nano funding for energy by the Department of Energy is about $10 million a year. This amount needs to be increased. Some energy companies are already getting involved. Halliburton, for example, has been interested in nano for at least a couple of years.

Most of the big U.S. energy companies are not yet turned on to nano, however. They haven't seen the handwriting on the wall as to their long-term survival as future energy providers, although some big petroleum companies are coming around. Some are currently buying solar companies to ensure their involvement in the solar revolution. Higher efficiency hybrid photovoltaic cells are being created, and new technology is helping lower energy costs per watt, increase the life and efficiency of solar modules, and meet green manufacturing guidelines.

Solar, though gaining ground in the United States, is especially strong in Europe. There are signs that European and other governments are investing in energy research, especially solar. In 2003, the Japanese government announced a $50 million program to certain energy industries focusing on nanotube production. That's a lot more funding pegged for energy than currently allocated in United States federal energy research budgets. Asia, however, is investing heavily for the long term, and if the United States wants to be in the energy leadership race, it needs to pick up the ball (or nanotube) and run with it.

Future of Energy

We're going to be using oil for the next 50 to 100 years. A lot of oil hasn't been recovered yet, but it's going to be extremely difficult and expensive to get at it. Nanotechnology discoveries are causing impacts across nearly every science and engineering field. As more and more nanoscale properties are understood, more powerful uses are being imagined. Perhaps the most globally exciting nano application is in the area of energy.

U.S. nanotechnology funding is spread broadly across many areas. Just over $1 billion is stretched thin (with only $10 million going toward energy). The Department of Energy is not putting much money into energy yet, although it has plans to ramp up.

As mentioned earlier, Nobel Prize winner Richard Smalley was a champion of alternative energy options, and he was convinced that new materials would be critical to being able to provide enough energy to go around. He believed that nanotechnology would help us meet global energy demands. Before his death in 2005, he wrote that by 2050, twice as much energy would be needed to fuel the Earth's energy needs. He believed that the Earth is "bathed in energy"—solar energy, nuclear energy, geothermal energy. He knew, however, that we don't have the technology to produce it, distribute it, or store it…yet.

Quiz

1. Geothermal reservoirs with temperatures of 180°C and higher are used to do all the following except

    (a) warm floors

    (b) heat buildings

    (c) wash and dry cars

    (d) power cities locally

2. In 2003, which government announced a $50 million program to certain energy industries focusing on nanotube production?

    (a) Africa

    (b) Canada

    (c) Germany

    (d) Japan

3. What is a common denominator of big problems facing humanity for the next 50 years?

   (a) no ice

   (b) energy

   (c) oceans

   (d) taxes

4. Which of the following is not one of the top 10 problems facing humanity?

   (a) energy

   (b) water

   (c) population

   (d) snoring

5. What is the projected amount of energy needed from a new clean energy source by 2050?

   (a) 3 terawatts

   (b) 6 terawatts

   (c) 10 terawatts

   (d) 15 terawatts

6. Of all the books currently published about fossil fuels, they nearly all agree that production will peak by

   (a) 2006

   (b) 2010

   (c) 2100

   (d) 2400

7. What energy source delivers a whopping 165,000 terawatts of energy per day?

   (a) sun

   (b) moon

   (c) philosopher's stone

   (d) natural gas

8. What nanomaterial has conductivity 10 times that of copper, 1/6 the weight, and is stronger than steel?

   (a) quantum wires

   (b) nanocrystals

   (c) buckyballs

   (d) space dots

9. According to Dr. Richard Smalley, how much energy will be needed by the middle of the twenty-first century?

   (a) same as today

   (b) 2 times

   (c) 3 times

   (d) 4 times

10. A terawatt (10^{12} watts) is roughly equivalent to

    (a) 1 hundred barrels of oil

    (b) 1 thousand barrels of oil

    (c) 10 thousand barrels of oil

    (d) 1 million barrels of oil

Part Three Test

1. Biosensors are important tools that provide selective identification of all of the following except

    (a) toxic chemicals

    (b) rock samples

    (c) biological samples

    (d) environmental samples

2. A silicon wafer is layered with a light-sensitive, liquid plastic called

    (a) goo

    (b) ticky tack

    (c) photoresist

    (d) photodesist

3. If cars were made out of advanced composite materials what percentage of gas could be saved?

   (a) 37%

   (b) 52%

   (c) 69%

   (d) 81%

4. What is the biggest problem with solar energy generation?

   (a) night

   (b) day

   (c) Saturdays

   (d) Sundays

5. When carbon is stored underground or in the deep oceans, it is known as carbon

   (a) paper

   (b) copy

   (c) black

   (d) sequestration

6. A carbon nanotube is stronger than

   (a) steel

   (b) copper

   (c) Kevlar

   (d) all of the above

7. Nano coatings on tennis balls have better seals that keep them from

   (a) bouncing

   (b) hitting the ground

   (c) losing air

   (d) getting lost

8. Extreme-ultraviolet light wavelengths are around

   (a) 1–5 nm

   (b) 10–15 nm

   (c) 50–100 nm

   (d) 193–248 nm

9. Materials science researchers have developed titanium dioxide nanotube arrays for generating

    (a) neon

    (b) palladium

    (c) hydrogen

    (d) sulfur

10. What bionanosensor molecule is used to detect blood glucose levels in diabetics?

    (a) benzene

    (b) ferrocene

    (c) aluminum

    (d) nitric oxide

11. Quantum dots confine electrons, holes, or electron-hole pairs to

    (a) water

    (b) donuts

    (c) cryogenics

    (d) zero dimensions

12. NCEM stands for the

    (a) New Car Engine Manual

    (b) No Catalysts or Electrophysical Methods

    (c) National Center for Electrical Maintenance

    (d) National Center for Electron Microscopy

13. The magnetic dipoles responsible for nanoring formation also create a collective magnetic condition known as

    (a) flux closure

    (b) plate tectonics

    (c) weddings

    (d) school lunch hours

14. The world's energy system needs to change from a mass-driven system (transporting coal, oil, and gas from place to place) to a

    (a) cryogenic system

    (b) wind system

    (c) foot powered system

    (d) transmission system

15. The terminals attached to the ends of the *n*-type channel of a transistor are called the source and the

    (a) train

    (b) mask

    (c) photoresist

    (d) drain

16. The following are all major problems facing humanity except

    (a) population

    (b) energy

    (c) food

    (d) keeping up with current fashion trends

17. In Shakespeare's time there were about 280 molecules of CO_2 out of every million molecules. Today, how many molecules per million are CO_2 molecules?

    (a) 220

    (b) 310

    (c) 380

    (d) 460

18. A de Broglie wavelength is

    (a) a tidal wave

    (b) the measure of wave movement (wavelength) of a particle

    (c) a new hairstyle

    (d) the length of a football field

19. A terawatt is equivalent to
    (a) 10^6 watts
    (b) 10^9 watts
    (c) 10^{10} watts
    (d) 10^{12} watts

20. Carbon sequestration is the method of
    (a) shaping raw diamonds into facets
    (b) storing carbon underground or in the deep oceans
    (c) hoarding coal for the winter
    (d) releasing carbon into the atmosphere

21. Nano machines that can carry out and analyze cellular processes normally done by biomolecules or entire groups of cells are components of
    (a) fiction
    (b) lithotripsy
    (c) cellular engineering
    (d) quantum optics

22. Regional energy blackouts are commonly the result of
    (a) late-night college parties
    (b) imbalances in electricity supply and demand
    (c) leaving all the lights on and running the dishwasher
    (d) poor quality light bulbs

23. Montmorillonite is a
    (a) rock formation in Nevada
    (b) highly ductile metal alloy
    (c) cotton ball
    (d) soft clay mineral of aluminum silicate that expands when it absorbs liquids

24. Transistors are the main parts of integrated circuits with huge numbers of transistors woven throughout and forming a
    (a) silicon chip
    (b) drain
    (c) mother board
    (d) source

25. The United States uses roughly how much energy yearly?

    (a) 400,000 watts

    (b) 850,000 watts

    (c) 3 terawatts

    (d) 10 terawatts

26. Researchers at the University of Nevada, Reno, Materials Nanotechnology Research Group have developed titanium dioxide nanotube arrays for

    (a) generating hydrogen by splitting water using sunlight

    (b) longer lasting cosmetics

    (c) industrial drilling bits

    (d) wheelchair wheels

27. Nanotechnology could boost the efficiency of today's solar cells of 20–30% to

    (a) 40%

    (b) 55%

    (c) 65%

    (d) 70%

28. The underground utility corridors are so full of copper wires in what city that it is said that it has the greatest amount of copper deposits in the world?

    (a) Atlanta

    (b) Houston

    (c) New York City

    (d) San Francisco

29. A lithography variation with liquid between the optics and the wafer surface is called

    (a) biomimetics

    (b) quantum mechanics

    (c) nanosurveillance

    (d) immersion lithography

30. The optical absorption of gold colloids produces a bright red color used in

    (a) cherry tomatoes

    (b) candles

    (c) home pregnancy tests

    (d) carpets

31. The 600 gigahertz speed barrier, invented by scientists at the University of Illinois at Champaign-Urbana, was broken using a

   (a) sling shot

   (b) bipolar transistor

   (c) laptop computer

   (d) off road motorcycle

32. It is estimated that the world's population will expand to how many people by 2050?

   (a) it will stay the same

   (b) 6–7 billion

   (c) 8–10 billion

   (d) 11–12 billion

33. When today's photovoltaic solar cells absorb a photon of sunlight, the energy gets converted to one electron, and the rest is lost as

   (a) color

   (b) foam

   (c) salt

   (d) heat

34. Through advanced composite materials, cars could save nearly what percentage of energy in the future?

   (a) 20%

   (b) 40%

   (c) 50%

   (d) 70%

35. Americans make up 5 percent of the world's population, but use what percentage of the world's energy?

   (a) 5%

   (b) 15%

   (c) 25%

   (d) 40%

36. The most commonly used methods to produce nanomaterials include all but the following

    (a) inert gas condensation

    (b) electrodeposition

    (c) lithification

    (d) sol-gel (colloidal) synthesis

37. When an electronics wafer is layered with a light-sensitive, liquid plastic, it is known as a

    (a) photoresist

    (b) masquerade

    (c) solvent

    (d) high end item

38. Gold nanorings have optical and electromagnetic properties that can be tuned by

    (a) a tiny switch

    (b) changing the ratio between the ring radius and wall thickness

    (c) very small technicians

    (d) changing the ring to Styrofoam

39. The sun delivers roughly how many terawatts of energy to the earth daily?

    (a) 10,000 terawatts

    (b) 50,000 terawatts

    (c) 138,000 terawatts

    (d) 165,000 terawatts

40. When nanosilica particles are mixed with polymer and salt in water (at room temperature), they

    (a) self-assemble into microcapsules

    (b) turn into solid ice

    (c) form sea stars

    (d) clump into gelatinous putty

PART FOUR

Future

CHAPTER 12

Business and Investing

Nanotechnology is a marketing firm's dream. Anything and everything called *nano* is hot these days. Look at Apple's new *iPod nano*. The pencil-thin, 3 ½ inch long, portable sound system, with a color display, up to 14 hours of battery life, and space for up to 1,000 skip-free songs, audio books and pod casts has lots of name recognition, but no real nanoparticles (some components are sized around 100 nm) inside.

In fact, so many current companies' products are sporting "nano names" that it's tough to separate the wheat (real nanotechnology products) from the chaff (nano hype). Some companies throw the term *nano* around like confetti, without really providing anything that qualifies as true nanotechnology.

Biotechnology's promising technology was a boon to the investment community, who happily bought stock in the rising star. But will newer, less-than-solid, faux nano business models and products mean investors will lose? We can hope that is not the case. Since the bottom line of most businesses is to create shareholder wealth, real monetary gains must result from patents and licensing of commercial methods and applications to support any company's growth.

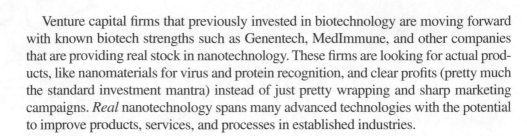

Venture capital firms that previously invested in biotechnology are moving forward with known biotech strengths such as Genentech, MedImmune, and other companies that are providing real stock in nanotechnology. These firms are looking for actual products, like nanomaterials for virus and protein recognition, and clear profits (pretty much the standard investment mantra) instead of just pretty wrapping and sharp marketing campaigns. *Real* nanotechnology spans many advanced technologies with the potential to improve products, services, and processes in established industries.

The Players

Almost 50 percent of the Dow Jones Industrial Average companies are either making or working on nano-related products. Industry giants such as Intel, IBM, Hewlett-Packard, DuPont, General Electric, Motorola, Sony, Siemens, and Xerox are just a few companies with nano interests.

A number of publicly traded nano companies could produce developments that would hugely affect their products and associated industries. These are commonly found in nanoelectronics, nanomaterials, and bionanotechnology.

With a growing number of companies (>350) in the United States and abroad that have their research and development sights set on the next big thing, nanotechnology will affect everything. The potential of building materials from components the size of a virus opens up tantalizing possibilities. Planes, trains, and automobiles, as well as biosensors, skin care products, computers, paints, and coatings are all at the brink of a technological tsunami that we can only estimate today. Many large companies (along with a lot of governments) are spending big money on nanotech research to keep from being left in the nano dust.

In 2004, estimates of more than $1 billion were spent by many of the world's governments in nanotechnology. It's possible that public and private interests could commit more than $2 billion each on nanotechnology in 2006. U.S. President George W. Bush mentioned nanotechnology in his 2006 state-of-the-union speech, and the administration's budget called for big dollars for physical sciences research, which would benefit nanotechnology. Bush's 2007 National Nanotechnology Initiative budget requests $1.275 billion distributed between 10 governmental agencies.

The lines between information technology, biotechnology, and nanotechnology are blurring as biology and technology appear to be merging in some areas (e.g., DNA chips). Or they could be considered as subsets of nanotechnology and its growing potential.

Companies such as Dow Chemical, ChevronTexaco, NEC, DuPont, ExxonMobil, and Mitsubishi Electric are investing in start-up nanotech companies founded by university faculty and/or teaming with companies that could feed materials to their processes. For example, Dow Chemical (with annual sales topping $33 billion), with customers and markets in more than 180 countries, joined with Starpharma of Melbourne, Australia, in an agreement with Dendritic NanoTechnologies, Inc., (DNT) for nano products using nanoscale polymers. This boosted DNT and Starpharma's patent portfolio in the field of *dendrimers,* a type of nanostructure with physical characteristics (branching) that makes them great tools to target diseases and deliver drugs to fight them.

DNT, with more than 30 patents in dendrimer science, sells and licenses more than 200 dendrimer types to pharmaceutical, diagnostic, and biotechnology companies. Currently, DNT is developing products for use with proteins, antibodies, anti-inflammatories, drug targeting proteins, and diagnostics for use with magnetic resonance imaging (MRI).

In 2004, Starpharma was the first company to begin human clinical trials of a dendrimer-based pharmaceutical for the prevention of human immunodeficiency virus (HIV) under the direction of the U.S./Food and Drug Administration Investigational New Drug application. The topical microbicide gel was developed as a medical preventative against the spread of HIV. Dow Chemical is also moving in dendrimer-based pharmaceutical applications.

STARTUPS

Even though the nanotechnology market is still very much "under construction," legitimate nano companies are coming up with great innovations and products. Many of these companies are evolving out of university research, headed by brilliant scientists who have never written a business plan. Some companies hire business managers to get things off the ground and make sure a reliable growth model is in place. Then, when everything is running smoothly, they hand off to operational managers and others to keep the forward momentum going. The most successful startups have a solid business plan from the beginning and stick to what they know. These companies pattern their growth after tried-and-true methods without getting carried away with overly grand (and unattainable) expectations.

The U.S. 2007 budget provides more than $1.2 billion for the multi-agency National Nanotechnology Initiative (NNI), bringing the total NNI budget since 2001 to more than $6.5 billion and nearly tripling the annual investment of the first year of the Initiative, according to the Office of Science and Technology Policy. However, products with some form of nanoparticle/nanomaterial components were worth around $26.5 billion in 2004, according to NanoMat, a materials research group of labs and companies with headquarters in Karlsruhe, Germany.

The nano-inclusive products on the market today include nanocatalytic chemicals, additives that keep paint mixed, sunscreens with nano zinc-oxide to protect against UV radiation, nanoparticle anti-scratch coatings for eyeglass lenses, anti-adhesive coatings for windows and car windshields, and frictionless lubricants to improve industrial tool workings and long-term use. Table 12-1 provides a sampling of several different nano-related companies. For a more comprehensive survey, refer to Appendix 2.

Table 12-1 Companies spanning every discipline offer nanomaterials, processes, or tools.

Company	Products
Acadia Research Corp.	Gene discovery, molecular characterization of disease
Altair Nanotechnologies Inc.	Lithium titanate spinel electrode nanomaterials
Applied Nanofluorescence, LLC	Optical instrumentation for nanotube study
Arryx, Inc.	Nano-tweezers to pick up and move nanoparticles
California Molecular Electronics Corp.	Invent, acquire, assimilate, and utilize intellectual property in the field of molecular electronics
Carbon Nanotechnologies, Inc.	Commercial production of carbon nanotubes (Buckytubes)
Cima Nanotech, Inc.	Fine, ultra-fine, and nanosized metal and alloy powders
Dendritech, Inc.	Dendrimer manufacturing and production
Dendritic NanoTechnologies Inc.	Branching molecules (dendrimers) with high application diversity (e.g., pharmaceuticals)
EnviroSystems	Hospital-grade, nanoemulsive disinfectant
eSpin Technologies, Inc.	Polymeric nanofiber manufacturing technology
Front Edge	Super thin rechargeable battery
Hysitron	Research/industrial instruments to measure nanoscale strength, elasticity, friction, wear, and adhesion
Intematix Corp.	Electronic materials; catalysts for fuel cell membranes
Kereos Inc.	Therapeutic nanoparticles/imaging tags to target disease
Lumera	Polymer materials and products
Mecular Electronics Corp.	Electronics and optoelectronic applications
Molecular Imprints	Nanoimprint tool maker for semiconductor and electronics industry
NanoDynamics	Nano silver, copper, nickel particles; nano-oxides; nanostructured carbons

Table 12-1   (continued)

Company	Products
Nano Electronics	New materials such as High K-Gate Dielectrics, metal gates, and silicides
NanoGram Corp.	Chemical compositions and slurries for computer chips
Nanohorizons	IP portfolio license from Penn State on manufacturing methods of thin-film nanostructures
NanoInk Inc	Anthrax detection
NanoOpto	Nanostructures for optical system building blocks
Nanophase Technologies	Preparation and commercial manufacturing of nanopowder metal oxides
Nanopoint	Allows scientists to peer inside/test living cells at resolutions of ≤ 50 nm. (e.g., IR, visible, and UV ranges)
NanoProducts,	Nanoscale powders, dispersions, and powder-based products, (single metal, multi-metal and doped oxides)
NanoSpectra Biosciences, Inc	Non-invasive medical therapies using nanoshell particles
Nanosphere	Analysis and ultra-sensitive detection of nucleic acids and proteins
Nanosys Inc.	Flexible/thin-film electronics, biologicals, and solar cells
Nano-Tex	Nanotech-enabled fabric enhancements/coatings
Nanotherapeutics, Inc.	Nanometer-scale particle delivery for pharmaceutical and over-the-counter products
Neo-Photonics Corp.	Nano-optical component maker
Novation Environmental Technologies	Iodine-based nanofiltration/disinfection of water purification
Ntera	Electronic ink and digital paper

Nanowires

Nanosys, Inc., a privately held company developing nanotechnology-enabled products in Palo Alto, California, uses nanowire technology developed by Harvard University chemist Charles M. Lieber. Lieber's nanowires are made from the same materials used for semiconductor lasers. With optical characteristics capable of detecting single molecules, nanowires may have an advantage over carbon nanotubes in the area of tiny optical sensors.

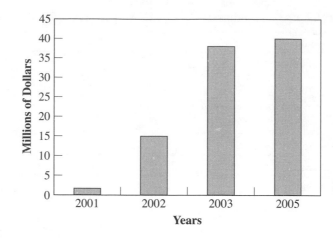

Figure 12-1 Nanosys is a nanotechnology success story with real products and sizable funding.

As described in Chapter 9, nanowires' potential for use in electronics and sensors looks good for a variety of uses. This fact has not been missed by the investment community. In November 2005, Nanosys, Inc., announced that it had raised nearly $40 million in private equity financing. The company plans to use the funding for continuing development and manufacturing scale-up of multiple products that include its proprietary, inorganic nanostructures.

Nanosys's product development plans include chemical analysis chips for pharmaceutical drug research, fuel cells for portable electronics, nanostructures for computer displays and sophisticated antennas, non-volatile memory for electronic devices, and solid-state lighting products. Nanosys's steep funding ramp up is shown in Figure 12-1. The company has also gotten millions of dollars in non-equity grants and contracts, as well as several joint development projects (e.g., multi-year development for nanotech-enabled displays with Sharp Corporation, Osaka, Japan).

Nanosys has one of the best technology lineups, with more than 400 patents and patent applications in basic nanotechnology areas. In 2005, Nanosys was listed as one of the top ten nano companies soon to go public by the *Journal of Nanotechnology Law and Business.*

Cell Targeting

As discussed in Chapter 7, buckyballs can be used to deliver a therapeutic drug to a specific target, minimizing side effects. They have been constructed into shapes that fit tightly into specific cell surface receptors. When treated with enzymes or proteins that disrupt a cell's reproductive cycle, buckyballs will serve as disease interceptors, similar to the human immune system. These nanotechnology treatments are being planned to treat patients with AIDS, cancer, arthritis, and other diseases.

Currently, sensitive microchips with intact DNA can pick up interactions between candidate antibiotics and target organisms. However, bionanotechnology would allow these biochips to be loaded with 100,000 times more chemical tests (with 100,000 times more sensitivity) using nanotubes.

Fluid Flow

The super small size of nanotubes allows them to pierce the skin painlessly (always a plus), especially for people who have to check their blood several times a day, such as Type II diabetes patients.

TheraFuse, Inc., is an emerging technology company that designs and develops infusion systems for delivering pharmaceutical and biopharmaceutical liquids. Since 2001, TheraFuse has created products that measure various fluid flow rates during therapeutic drug infusion (e.g., very low insulin infusion rates for newborns to very high infusion rates needed for clinical intravenous drug delivery). Currently, the Vista, California, company is creating a skin patch for diabetic use. By using nanostraws to draw blood and check glucose levels and then inject insulin when needed, the diabetic patch will make the mechanics of the disease a lot easier, especially for children.

BioHazards

Nanotech offers incredibly sensitive systems for detecting biohazards. Anthrax threats could easily be detected and dealt with using nanotechnology methods able to recognize an anthrax spore's surface. Early methods required technicians to crack a suspect spore, release it, and check its DNA. This process took a lot of time.

Chicago's NanoInk, Inc., has licensed technology to test anthrax spores by using a super small "pen" that writes nanoscale lines. By programming a computer to draw thousands of patterns based on unique, anthrax spore surface features, patterns that bind only to anthrax are discovered. NanoInk then reproduces these for use in detection kits.

Computers

Computer and chip companies who were among the first nanotech visionaries to dig down to the nanoscale world are still big investors.

Laptops and palm-sized computers will perform better with nanotechnology components. Batteries with carbon nanotubes and nanoscale lithium particles will recharge faster, last twice as long, and maintain higher energy densities. Since

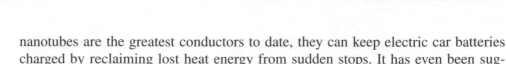
nanotubes are the greatest conductors to date, they can keep electric car batteries charged by reclaiming lost heat energy from sudden stops. It has even been suggested that gas tank materials using nanotubes could store hydrogen for fuel cell–powered cars. Fossil fuels and their associated problems would no longer be a major part of the transportation picture.

It's a fairly safe bet that silicon chips will keep getting better for at least another 5 to 10 years. Since the semiconductor industry has invested billions of dollars in silicon companies, nanochips may not become economically big until silicon hits the wall. It's also true that scientists/engineers still need to discover manufacturing methods that can direct nanotubes to self-assemble into intricate circuit designs consistently.

Assemblers

One engineering vision is of an *assembler*, first described by Eric Drexler, nanotech pioneer and head of the Foresight Institute in Palo Alto, California. A nanoscale robot that could be programmed to assemble atoms into gears and other nanomachine components would be a shortcut to more complex (if that's possible) and time-consuming methods. As we discussed in chapter 3, that vision is largely science fiction, but if assemblers could be made, they could possibly meet a lot of our material needs. From dust bunnies and dirt molecules, they would collect atoms to assemble computers, appliances, and whatever met consumer demand . We can always dream....

NanoBusiness Alliance

Science and engineering researchers are not usually in the business of, well, business. They are brilliant, extremely focused people who come up with things that the rest of us can't imagine, let alone identify, characterize, or manufacture. For this reason, some researchers need guidance moving from their labs to startups.

Similarly, entrepreneurs could use some help in understanding the complexities of nanotechnology. To direct and advise both groups, the NanoBusiness Alliance was formed in 2004 by multiple stakeholder groups, Fortune 500 companies, universities, startups, and public/private partnerships. To date, the alliance has more than 200 members looking toward huge financial rewards. The fact that the National Science Foundation projects nanotechnology as a $1 trillion market by 2015 has not been lost on forward-thinking leadership and members of the NanoBusiness Alliance.

Executive Director Sean Murdock stated before the research subcommittee of the U.S. House of Representatives Science Committee in June 2005 that alliance members believe that nanotechnology will be one of the key drivers of business success, economic growth, and quality-of-life improvements of the twenty-first century. Murdock called on the U.S. government to be the gold standard for commercializing nanotech innovations.

A company that is taking Murdock's charge to heart is Lux Research/Capital, a research-driven investment company established in 2000 and focused on nanotechnology. Lux Capital has been watching nanotechnology evolution carefully. The company is going deeper than catchy nano branding and looking for real nanomaterials in improved products.

Lux Capital and other investment firms believe that nanotech has begun what might be compared to the California gold rush days. The big difference being, in time, that the payoff of true nano companies should be much better than finding gold dust in a stream. However, plenty of uncertainty exists as to whether most nano startups will boom or bust. Even with the amazing progress being made by companies in developing nanoscale transistors and prototype nanotube circuits, their payoff may still be years off.

Nanotechnology is also hugely multi-disciplinary. Instead of one or two people making the lion's share of breakthroughs, chemists are working with physicists and biologists who might consult materials scientists or computer scientists using super computers to model very specific chemical/physical interactions. This once rare interconnectedness is a very big part of nanotechnology development. In fact, many business and research people believe it is essential.

Implementation 101

Historically, improved instruments have been crucial to the progress of science and technology. In fact, some people believe that for nanotechnology's promise to come true, a big investment jump in the development of simple, accurate, and reliable nano tools is critical.

University, industry, and national laboratory collaboration in developing and commercializing nano instruments and specific tools is also important. When analytical facilities such as high-performance beam sources (e.g., electron beams) are available to provide affordable, state-of-the-art capabilities to the research and engineering communities, then commercial projects will be more able to see the light of day.

At the nanoscale, instrumentation gives us the ability to see, touch, move and build nanoparticles/nanomaterials. Supplying researchers with the tools to discover and study new chemical, physical, and biological happenings in the nano world is an important area for investors to consider. Early instrumentation investment are likely to yield benefits in all aspects of nanoscale science and technology.

What to Watch For

Several examples of new instruments and their nanoscale uses include nanoparticle and nanotube manipulation, near-field scanning optical microscopy, and surface force microscopy.

SINGLE MOLECULE MANIPULATION AND MEASUREMENT

Tools to move and measure single-molecule properties make it possible to check out nanomaterials' important capabilities. As described in Chapters 5 and 6, biology and medical care will benefit if clinicians can influence molecular chemical/physical operations at the nanoscale level that make up cells, tissues, and organs. Structural polymers, adsorbents, and catalysts (e.g., proteins and enzymes) may become so specific that they could be adjusted to meet an individual patient's medical needs.

In the past, molecular measurements were mostly averaged calculations, since scientists weren't able to test a single molecule's behavior. While averaging techniques work, they hide individual peculiarities that are needed for researchers to understand completely the oddities of molecular behavior. Ground-breaking instrumentation gives them the tools to explore the largely unknown single molecule world.

Some of these materials and methods (described in other chapters) are summarized here:

- Unique electrical and mechanical properties of carbon nanotubes have been proven through individual nanotube measurements.
- Molecular displacement measurements are now possible (i.e., recognition between antibody/antigens or complementary DNA strands).
- Natural molecular motors are responsible for DNA transcription, cellular transport, and muscle contraction.
- Optical tweezers provide direct molecular measurement and manipulation of protein folding configurations and dynamics.

New and Improved Tools

Because nanomaterial manipulation and measurement tools often come out of university and government laboratories, industry keeps a close eye on new discoveries and invests in and commercializes those that look the best.

This was demonstrated in the university/industry relationship between Washington University in St. Louis, Missouri, and Zyvex Corporation. Their collaboration has resulted in a new tool for manipulating and imaging (scanning electron microscope) nanoscale objects simultaneously. This instrument permits researchers to pull, bend, twist, and buckle nanotubes into 3D configurations. The low-cost manipulator offers good movement, precision, small size, and quick assembly of nano components.

Another product enhancement story involves EnviroSystems of San Jose, California, which "nano-sized" regular disinfectant and made it better. EnviroSystems' non-irritating, hospital-grade disinfectant cleans as well as standard cleaner/bleach disinfectants, but it doesn't need U.S. Environmental Protection Agency special handling or warning information. This currently makes it the only non-corrosive, non-irritating disinfectant on the market without warning labels.

The company's product, EcoTru ®, uses nanoscale emulsive particles that are so small they can penetrate bacterial cell walls and destroy them from within. It's like a nanotech smart bomb (i.e., particles target bacteria and germs, while leaving human/animal cells alone) for use against microorganisms. In addition, the formula isn't toxic. In fact, it is advertised to hospitals, dental and medical clinics, home care, and laboratories with the following claims:

- 100 percent effective against HIV, *Staphylococcus, E. coli* and many other viruses, bacteria, and fungi
- Kills *Tuberculosis* B in 5 minutes
- No protective clothing required
- No toxic fumes
- Non-flammable
- Cleans and disinfects in one step

Through nanotechnology, EnviroSystems made an established product work a lot better on surfaces such as metals, plastics, synthetics, rubber, glass, and painted surfaces. Since the product is non-corrosive, it can be used on delicate equipment and surfaces as well. In fact, when used in Africa as a preoperative antiseptic for surgeries, the product got rid of post-operative infections in every patient it was used on (500 out of 500). Although it must pass FDA testing before being used in U.S. hospitals, these early results have made clinicians (always looking for safe ways to kill bacteria/viruses) sit up and take notice.

By changing a few molecules of a standard industrial disinfectant, EnviroSystems has developed an improved antibacterial cleanser. This kind of existing product nano overhaul makes the development of important new products possible.

Local Nano Hopes

Various state governors, university presidents, and entrepreneurs are trying to figure out how to turn their home turf into the next U.S. nano-electronics capital. Silicon Valley in California has a home court advantage, but in Texas, Austin's strong research facilities at the University of Texas and Houston's nano force at Rice University have given weight to a number of startups.

In upstate New York, the city of Albany is a new nano-electronics player. After receiving more than $7.25 billion in private investments, much of it highlighting the semiconductor industry, nanomaterials have become a bit hit.

For example, the Dutch chip-lithography giant ASML Holding (a market leader in immersion technology), along with IBM, is building a new $400 million research center in Albany. It will be ASML's first large-scale, non-European research and manufacturing facility. This follows on the heels of Albany's 2002 successful courting of SEMATECH International, a leading semiconductor research consortium. SEMATECH built its new $403 million research center in upstate New York, with the state throwing in around $75 million.

International Outlook

If you think nanotechnology is just for scientists and engineers, think again. Nanotechnology is a forward-thinkers investment, not a sci-fi fad that will disappear overnight. Nanotechnology's potential impact in dozens of fields makes it, as you might expect, a hot international topic.

EUROPEAN UNION EYES NANO

In 2004, European technology companies such as Philips, Nokia, Ericsson, AMD, and IBM decided that if Europe wanted to lead the world in the next "oil of the future," it would need to invest at least 6 billion Euros per year to switch from micro to nano-scale electronics. The *European Nano-Electronics Initiative Advisory Council,* a public-private partnership charged to come up with and implement a European nano-electronics research agenda, first met in 2004.

Some of its goals are listed here:

- Support and contribute to European/private research and development of nano-electronic investments
- Speed up innovation and the use of research technologies
- Improve EU nanoelectronics competitiveness and productivity
- Get rid of obstacles to coordination and facilitate/accelerate market penetration of new technologies
- Balance innovative and policy oriented research, while aligning research/technology with European policies and regulatory frameworks
- Make the EU more attractive to researchers and industry

- Increase public awareness, understanding, and acceptance of nanotechnologies

Only time will tell whether they will be successful in snagging important nanotechnology products and methodologies before the rest of the world.

Biotech to Nanotech

The European Network of Excellence program, (Nano2Life) with nearly 200 scientists, 23 research organizations, and 12 countries, has joined with industrial partners to identify regional centers, disciplines, and expertise available for collaboration. The Nano2Life program has been funded from 2004–2008 and is planned eventually to become the European Institute of Nanobiotechnology, a long-term European science/ engineering research institute. Its main stated goal includes developing joint research projects on four major technical platforms: functionalization, handling, detection, and integration of nanodevices. In fact, since April 2004, more than 30 projects have been established with 20 collaborative research projects active and funded by the European Community.

The centers participating in the Nano2Life collaborative research projects and workshops are from Denmark, Germany, Greece, France, Sweden, Israel, and Australia to name a few. Spin-offs and startups are also intended to come about through Nano2Life collaborations.

Nano Forecasts

In a 2005 report by Lux Research, "The Truth About Nanotech Tools," market sizing and forecast numbers are provided for all three nanotech tool areas. In confidential CEO and marketing executive interviews with 21 global tool vendors and research leaders at 49 corporations, universities, startups, and other U.S. nanotechnology companies, Lux Research tested the nanotechnology industry.

Interview data was meshed with forecasting models of individual tool types in different areas. The results were also subjected to a peer review process with nanotech tool vendors and top industry people. Figure 12-2 illustrates how tools and various components fit into the overall nanotechnology development picture.

The 2005 market for nanotechnology tools was mostly composed of inspection tools (about 95 percent of 2004 revenue) from the few years of early nanotech explosion. This was around the same time that several university nanoscience/nanotechnology centers were built. In fact, some experts believe the nano tool market has become saturated.

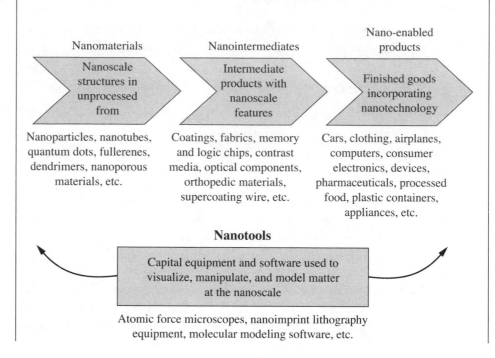

Figure 12-2   Nanotechnology includes the interplay between many technical areas and products.

The Lux report explains that faster and cheaper tool alternatives are available for routine quality control of new nanomaterials now, but that nanotechnology manufacturers are getting to the point that they need only inspection tools like scanning probe microscopes and electron microscopes for methods development and spot checks. It's possible that as production matures, the number of new research tools needed will drop.

However, it's important to point out that future nanoscale science and technology applications/products can be sped along by nanoscale measurement tools with *improved* capability. By expanding instrument capabilities (to describe new phenomena and fully describe nanoparticles/materials), nanoscience advances will help healthcare, electronics, environmental safety, law enforcement, and national security applications leap ahead.

Worth Watching

The public television stations market their programs as "TV worth watching." Nanotechnology is science/engineering worth watching. Nanotechnology progress has the potential to be huge for two reasons. First, it is expected to provide $1 trillion in revenues by 2015. Second, it will impact virtually everything from personal items (e.g., cosmetics, sunscreen, and clothing) and medical care (wound healing and cell targeting) to energy transmission and fuel cell materials.

Some experts see nanomaterials as *the* investment opportunity field of the next 10 years. Although medical applications in several areas look good, other investors are wary of the many FDA regulatory hurdles that must be overcome before getting to market. However, since nanomaterials' special properties have the potential to impact many, many market goods, stay tuned.

Quiz

1. What percentage of Dow Jones Industrial Average companies either make or are working on nano-related products?

    (a) 25%

    (b) 40%

    (c) 50%

    (d) 62%

2. Publicly traded nano companies are focused mainly in the areas of nanoelectronics, nanomaterials, and

    (a) oil refinement

    (b) marine geology

    (c) bionanotechnology

    (d) diet drinks

3. Denmark, Germany, Greece, Sweden, Israel, and Australia are among the countries working on nanotechnology research in a collaborative effort called

    (a) NanoNow

    (b) Nano or Bust

    (c) Living the Nano Life

    (d) Nano2Life

4. What popular product is using nano as a big marketing tool?

   (a) super chewy nano bubble gum

   (b) nano cocktail napkins

   (c) iPod nano

   (d) nano nose rings

5. Nanotechnology is expected to provide how much in revenues by 2015?

   (a) $100 million

   (b) $500 million

   (c) $800 million

   (d) $1 trillion

6. The 2005 market for nanotechnology tools was mostly composed of

   (a) dental tools

   (b) inspection tools

   (c) toasters

   (d) tiny toy robots

7. Formed in 2004, the NanoBusiness Alliance has approximately how many members?

   (a) 100

   (b) 165

   (c) 200

   (d) 231

8. Nanowires' ability to detect single molecules has great potential in the development of

   (a) optical sensors

   (b) anti-aging cream

   (c) magnetic bowling pins

   (d) waterproof swim suits

9. Investors are looking for

   (a) decreasing profits

   (b) actual nanotechnology products

   (c) ways to replicate the dot.com bubble

   (d) better airline food

10. Nano tools help researchers isolate, understand, and use motors as

    (a) actuators

    (b) cursors

    (c) carbonators

    (d) innovators

CHAPTER 13

Nanotoxicity and Public Policy

Killer organisms, doomsday weapons, evil geniuses, and science run amok are often the stuff of science fiction thrillers. Bigger and fiercer monsters and machinery grabs audiences' attention and pocketbooks. Today's media types unveil the latest sci-fi hero or evil empire with everything from action figures to video games. Nearly everything seems capable of destroying the world!

Over time, the public has been trained to look for the "dark side" of technology. With lawsuits focusing on everything from too hot coffee to dangerous pharmaceuticals, the public questions everything and distrusts everyone (until it has proof to the contrary).

The media often champions this approach, and with enough spin, anything becomes the big story. It's not surprising, then, that every new technology to come along is viewed with a mixture of unease and excitement.

However, communications research has shown that the media may not have as much of an impact on people's opinions about technology as previously thought.

A lot of nano public opinion research shows that people are generally accepting of and positive about technology. Although they may be fairly unfamiliar with nanotechnology, most people are taking a wait and see attitude. As in the past, some new materials have seemed promising, only to cause unanticipated harm in the long run. Discoveries can be a two-edged sword—they allow progress, but they may open a Pandora's Box if not studied carefully.

Just because something is useful doesn't mean it is risk-free. It wasn't too many years ago that everyone was praising asbestos, thalidomide, chlorofluorocarbons, and DDT as important advances/products in many areas. Only after big problems showed up did we think that increased caution might be important.

Some critics think that the same technologies that many scientists believe may someday cure cancer and solve our global energy transmission problems could also destroy the planet. They want all advanced bio and nano-research stopped to keep it from wiping out the human race. Well, who wouldn't want a global marauder stopped? Time will tell if nanotechnology is a boon or bane to society.

In an April 2000 article in *Wired*, "Why the Future Doesn't Need Us," Bill Joy, co-founder and chief scientist at Sun Microsystems, stated the current pace of technological progress was a very real threat to the future of humanity. He saw three big threats to humanity: genetic engineering, nanotechnology, and robotics.

Although no one has a crystal ball to see the future, world collapse predictions seem a bit over-reactive. While it's true that science/engineering methods used for ill intent are dangerous, to give up on good things out of fear seems counterproductive. The watch word must be thoughtful development, education, standardization, and international cooperation. Fifty years ago, many people thought the world would soon end in nuclear holocaust, but we still survive, even though the risk remains real.

It's important to heed concerns (part of the scientific method is to question everything) and go forward with caution and care. At the same time, we need to realize that the best could be just beyond our reach.

Nanotechnology and You

The industrial age has provided us with new products galore. New discoveries and materials have made it possible for old products that were originally made of wood and paper to be updated with oil and petroleum-based materials. A simple example is restaurant take-out boxes; first made of paper, many are now plastic, which don't get soggy and leak. Plastic, based on petroleum, is a technological product.

Science, engineering, and computer design were the glory fields of the twenty-first century. There was only one problem, however: sometimes they leapt before they looked. New was cool. Cutting-edge was the place everyone wanted to be.

We weren't as careful with or didn't care about new products' impacts. As long as something was "new and improved," it was seen as good!

Eventually our carelessness came back to bite us. The air, land, and water became polluted, often with a mixture of compounds. The United States and other governments were forced to spend billions of dollars cleaning up chemical runoff and toxic landfills, and educating industries and the public on pollution. It took a while, but we are slowly starting to realize that the Earth can't put up with poor stewardship forever.

With nanotechnology, more and more scientists, government officials, and industrialists are trying to look before they leap and do things right from the start. Everything nano is promising news today, but by learning from past oversights, we can identify and eliminate potential toxic effects from the beginning.

Solubility and Toxicity

In early assessments of emerging nanotechnology, the solubility of nanoparticles has been of particular interest since soluble molecules can enter biological systems and the environment more easily.

You learned in Chapter 1 that the first fullerene was Buckminster fullerene (C_{60}). Since its original discovery in 1985, research has found that C_{60} molecules cluster into nanoscale crystals that are active in watery environments. Additionally, soluble C_{60} clusters have been shown to be *cytotoxic* (cell killers) *in vitro* and have been linked to brain tissue damage in fish.

However, Rice University scientists Vicki Colvin, Jennifer West, and Joe Hughes discovered that by attaching small molecular fragments to the surface of C_{60} molecules, in a process called *functionalization*, the cytotoxic effect was eliminated. Non-functionalized fullerenes (used in the fish study) were toxic to 50 percent of the cultured cells at a concentration of 20 ppb. Figure 13-1 illustrates this neutralization of toxicity.

The functionalization method seems to work additively. By attaching more and more molecular fragments to the C_{60} molecules' surfaces, the cytotoxicity drops until the fullerene's surface is covered and not enough nano-C_{60} exists to kill 50 percent of the cells. This important study showed how nanoparticle risks are dependent on the material's particular form and that simple techniques can be used to reduce and/or eliminate risks.

So the take home message was that once the nanomolecules' surface properties were better understood, it took a relatively simple chemical modification to negate risks. This may or may not work for other nanoparticles, materials, devices, and processes. Nanomaterials' unique properties (e.g., surface chemistry, strength, reac-

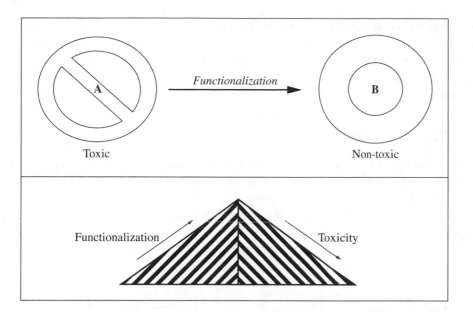

Figure 13-1   Nanotoxicity goes down as functionalization increases.

tivity, and thermal/electrical conductivity) may require other application-specific changes before their fullest potential can be utilized.

Moreover, modifications often affect a material's properties. In some applications, modifying the surfaces may engineer out the very property that is desired. However, innovations in communications, medicine, materials, energy production/efficiency, water treatment, and environmental clean-up may become reality.

NANO-CELL INTERACTIONS

Understanding the toxicological effects of engineered nanoparticles in liquid phase reactions is an important goal of some nanotechnology research areas. While a lot of information is available about aerosol effects upon living systems through inhalation exposures, very little is known about the toxicology of engineered nanoparticles in the liquid phase.

Toxicology studies on engineered nanoparticles' interactions with cells are underway to determine whether surface coatings can be used to affect the toxicity of the nanoparticles. Testing has shown that in water, C_{60} in concentrations up to 100 ppm forms nanoscale clumps (10–50 ppm) with hydrophilic (dissolves in water) surfaces. This is very important because C_{60} itself is considered insoluble (hydrophobic), so the clustering makes it soluble and environmentally significant, for example, if the particles were to be released into a natural stream or river system.

Other nanoparticles have served as bactericidal agents against both gram positive and negative bacteria. Contact between nanoparticles and bacterial membranes points toward possible disinfectant effects in a hospital setting, since nanoparticles' have a tendency to become airborne (super small size) and/or cross biological membranes.

Derivatization

Size, though, isn't the only factor for predicting the nanoparticle toxicity: chemical composition, aggregation, surface charge, particle shape, structure, and surface properties/coatings are among the other factors that must be considered.

In C_{60} cytotoxicology tests using human dermal (skin) cells, the toxicology of nano-C_{60} in culture was 20 ppb. Although this is fairly toxic compared to other cytotoxic chemicals, the toxicity of nano-C_{60} and other water soluble fullerenes was connected to their level of *derivatization*.

> *Derivatization* takes place when a chemical compound is transformed into a product of similar chemical structure called a **derivative**.

Often, a chemical compound's functional group (a side element or molecule that helps a compound function) is derivatized in a reaction. This changes the original compound (A) into a product (B) with a slightly different reactivity, boiling point, melting point, solubility, and/or chemical composition. Derivatization is also important in altering unwanted properties (such as toxicity) in a biological or manufacturing process.

Fullerene toxicity affects cell membranes through the production of oxygen radicals. Tests show that this toxicity can be shut down through the derivatization of the C_{60} molecule.

Work with nanoscale quartz, titanium and labeled iron oxides is being done in cell culture and lung studies to figure out how specific particle size and surface chemistry change nanoparticles' biological effects/biodistribution.

MOBILITY

Another globally important angle of nanotechnology risk concerns the extreme mobility of nanoparticles. Because they are super small, they can get into almost everything. Think of dust that is so fine, it can't be seen. This is a benefit when attacking disease and local viruses, but it could cause a lot of environmental problems if not handled correctly.

In Chapter 7 you saw how iron may be used to clean up environmental contaminants. Before this method and others can be put to widespread use, researchers must be sure that they can recover the nano-iron after the toxic clean-up.

However, this might not be necessary. Wei-xian Zhang at Lehigh University claims that because iron is already found in the environment, these particles may not need to be recovered. However, the same may not be true for particles of different composition. Since different nanoparticles can be highly reactive with certain compounds in certain conditions, any broad environmental use, no matter how safe it seems at the time, must be handled carefully.

In order for researchers, industry, and the public to buy into the awesome potential that nanotechnology offers, we have to learn from our mistakes with DDT and other "helpful" chemicals. This is being done through a variety of efforts currently underway in the United States and abroad.

Icon

The International Council on Nanotechnology (ICON) was established at Rice University to assess, communicate, and reduce environmental and health risks connected with nanotechnology. To carry out this goal, ICON is working with academics, industry, government officials, and representatives of environmental organizations. As described in Chapter 7, its activities include research into nanoparticle/cell interactions, policy, standards, terminology, and social analysis on risk perception and education.

NANOMATERIALS DATABASE

By combining the nanotechnology industry's resources, along with governments, and academia, ICON can network related projects and serve as a central repository of information. Along with the Department of Energy and scientists from Rice University and Oak Ridge National Laboratory, a new database was created that is centered on the environmental, health, and safety implications of nanomaterials. This catalog of scientific literature (http://icon.rice.edu/research.cfm) was launched in August 2005 to help researchers and government agencies make up-to-date decisions about nanomaterials' safety.

The CBEN database, free of charge on the Internet, is an evolving document that is as useful for the general public as for researchers. Visitors are able to tailor searches to "nanoparticle type" and "production method." Information on the latest information about health and environmental implications of nanomaterials can be found on this website.

Future ICON plans include summaries, written for the general public, of the database's most significant papers. These would include review articles such as the health, safety, and environmental impact of cadmium selenide quantum dots or the pulmonary toxicology of engineered nanoparticles.

ICON's goal is to create an information bank that will keep everyone current on the latest nanoscience/nanotechnology discoveries and public policies. In this way, innovations, standards and regulations can be organized.

Responsible Development

The National Nanotechnology Initiative's (NNI) plan for responsible nanotechnology development is divided into several areas. These include environment, health, and safety implications as well as ethical, legal, and other public issues. Realizing that new science/technology advances offer public benefits as well as risks, the NNI has begun researching these priority areas.

Federal regulatory and research organizations are working with the NNI to identify existing regulatory procedures that cover the production and use of nano-materials. Where regulation gaps exist, steps will be taken to change or increase regulatory standards.

BEST PRACTICES

The National Institute of Occupational Safety and Health plans to develop and issue a best practices document for working with nanomaterials. It will also work with industry and academia to develop nanomaterials nomenclature.

As discussed in chapter 2 nanotechnology academic and industry members are working toward the standardization of measurements. Early standardization will allow scientists to compare apples to apples. Best work practices will lay out a plan to ensure the safety of technicians and those working in all parts of the nanotechnology research environment. Once these practices are well established and a size-able history of well-investigated nanotechnology uses in several key areas (e.g. medical, environmental, health and safety, etc.) is completed, public and private concerns should drop off quite a bit.

SOCIETAL OUTLOOK

Levelheaded nanotechnology development requires that governments consider and address societal outlooks as well as science/engineering challenges. Some of the different angles to think about include the following:

- Access to nanotechnology benefits
- Effects on jobs
- Medical innovations and methods
- Manufacturing impacts
- Potential health/environmental effects
- Privacy issues (e.g., information from nanosensors)

Responsible development of nanotechnology also means that the government has to establish public communication channels through the NNI. Since technology has gotten so much attention in science fiction and the movies, it is understandable that people are a bit skittish about science that seems too good to be true. Open information lines allow the public and the government to make well-informed decisions and build a solid knowledge/trust foundation.

There's an old saying: "Perception is truth." In our fast paced world, everyone is bombarded with information, opinions, research data, and ideas. It is often tough to tell the difference between truth and what is thought to be true (but based on rumor or appearances). Understanding and acceptance of new technology are key parts to weaving new methods and materials into the fabric of everyday life.

GRADIENT INTRODUCTION

Before the government can "get it right" with nanotechnology, it needs to promote a graded/gradual introduction of nanotechnology into common use. This is not intended to slow nanotechnology's introduction, but to make sure all issues are considered. Such a plan could use existing abilities and structure to make sure that nanomaterial risks are identified before they are included in manufactured products; manage identified risks from start to finish, keeping safety of workers, users, and the environment in mind; develop research and use standards that make sense; and include everyone in the process (e.g., industries, health organizations, consumer groups, environment organizations, investors, and the public).

Environment, Health, and Safety Implications

The NNI sponsors a wide range of research to evaluate the environmental, health, and safety impacts of nanotechnology. The NNI's research support has grown with the discovery of new nanoparticles, nanostructures, nanomaterials, and nanotechnology products.

As both nanotechnology's cheerleader and watchdog, the NNI has a tough, but interesting job.

The NNI's various policy projects include the following:

- Study potential health risks of nanomaterials.
- Integrate the efforts of the National Institute of Environmental Health Sciences, National Institute of Occupational Safety and Health (NIOSH); Environmental Protection Agency; Department of Defense; Department of Energy; and National Science Foundation.
- Develop new standards with the National Institute of Standards.
- Facilitate communication among the member agencies.
- Identify and rank research necessary for regulatory decision-making.
- Encourage better interaction within governmental, industrial, and scientists/ engineers at colleges and universities.

These NNI tasks are coordinated by the Nanotechnology Environmental and Health Implications Working Group (NEHIWG), with supporting membership from nanotechnology research and regulatory agencies.

One of the working group's aims is to understand the fate and transport of manufactured nanomaterials along with the development of nanomaterial life cycle evaluation methods. Research focusing on nanotechnology's environmental and health implications will grow as needed. Working with the international scientific community on important nanotechnology research is important for building a global picture of nanotechnology use and focus.

CURRENT NNI RESEARCH

In 2004, the National Nanotechnology Initiative funding for initial health and environmental research was estimated at $105.8 million, or around 11 percent of total NNI funding. This amount included basic research, applications, and implications of nanoscale materials. The NNI's federal research and development program that's coordinating nanotech research from 23 separate U.S. government agencies has earmarked another $39 million to study the health and environmental effects of nanotechnology in 2006.

To understand the health and environmental impacts of molecular manufacturing, research into naturally occurring nanoscale material exposures is being considered. Currently this type of molecular exposure comes from desert dust, volcanic ash, forest fire smoke, bacteria, and viruses. Older technologies and processes produce substances such as combustion soot, diesel exhaust, paint pigments, and welding fumes.

The important thing to remember about nanoparticle risk is that it is related to the level of toxicity and length of exposure time.

$$\boxed{\text{RISK} = \text{TOXICITY} \times \text{EXPOSURE TIME}}$$

For example, a person's risk is affected by a material's toxicity just as much as the time he/she is exposed. If someone is exposed to a mildly toxic compound for a short time, then the risk is fairly low (depending on the compound and tissues affected), but if exposure extends too many years, risk goes up.

$$\uparrow \text{RISK} = \uparrow \text{TOXICITY} \times \uparrow \text{EXPOSURE TIME}$$

$$\uparrow \text{RISK} = \Downarrow \text{TOXICITY} \times \uparrow \text{EXPOSURE TIME}$$

$$\Downarrow \text{RISK} = \Downarrow \text{TOXICITY} \times \Downarrow \text{EXPOSURE TIME}$$

Research geared toward a better understanding of nanoparticles' impact on human health and the environment will help to prevent and clean up potential problems. Although exposure to engineered nanomaterials such as carbon nanotubes or buckyballs is currently limited to people who work with them in laboratories and production facilities, that may change a lot in the next decade.

Risk Assessment

Manufacturing exposures and risk must be kept in perspective. Engineering Professor Mark Wiesner at Duke University and colleagues are studying various nanomaterials for toxicity problems.

They performed a comparative risk assessment for industrial fabrication of different nanomaterials. Based on the likelihood for large-scale production and commercialization, five nanomaterials were tested: single walled carbon nanotubes, buckyballs (C_{60}), quantum dots, alumoxane nanoparticles, and nano-titanium dioxide. The study looked at risks associated with nanomaterial fabrication, not specific impacts or risks of the nanomaterials themselves.

The researchers kept track of input/output materials, as well as waste streams for each fabrication step. This information was then entered into a database that included factors such as temperature and pressure. The physical/chemical properties and quantities of the test materials were used to compare relative risk based on volatility, toxicity, flammability, mobility, and persistence.

Test results show that fabrication risks are fairly comparable to standard manufacturing risks. Figure 13-2 shows how normal operational risk of manufactured nanomaterials compares to operational risks in other industries.

Products like cosmetics and sunscreens among others, are also on the market that may need a closer look as long term exposure data becomes available.

NNI-funded research is intended to boost basic understanding of nanomaterial interactions at the molecular and cellular level through *in vitro* and in *vivo* experiments and models; enlarge knowledge of nanomaterials' interactions with the environment; enhance understanding of the fate, transport, and transformation of nanoscale materials in the environment; identify and characterize possible exposure; determine possible human health impact; develop standardized control

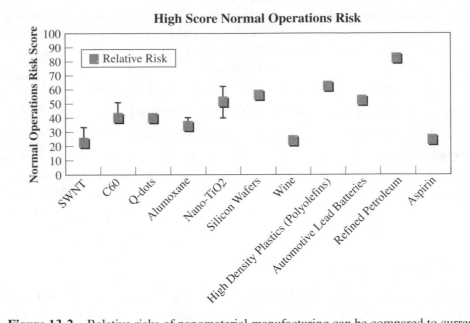

Figure 13-2   Relative risks of nanomaterial manufacturing can be compared to current material manufacturing risks.

methods for occupational exposure to nanomaterials; and establish safe working procedures for nanomaterials.

In collaboration with the NEHIWG, NIOSH's new guidelines for best practices will be a huge step toward safe working conditions when producing and handling nanoscale materials. NIOSH and other NNI departments and agencies plan to deliver these guidelines to the research, business, emergency response, and related communities and outline them on the NNI and NIOSH websites.

Getting the Word Out

In order for nanoscience/nanotechnology to be explained simply, information must be shared with the public and the media in a way that makes sense. Discoveries, technologies, and applications must be separated from media hype. It doesn't fly if nanotechnology is seen as science fiction—fantastic or frightening technology. There is enough amazing real science/technology out there without adding to the confusion with way out claims for nanotechnology.

Nanotechnology must be generally understood to the extent that the workings of the telephone, computer, or a vaccine are understood. People don't want to become experts. In fact, most people don't really care how things work until they break down.

A lot of people have the *Jerry Maguire* film attitude, "Show me the money!" They want to know what nanotechnology has done for them lately. What will it accomplish in the future? People want a general understanding of the world and how new technology affects them personally.

NANOTECHNOLOGY EDUCATION

Nanotechnology touches society in many areas—economic, cultural, ethical, and legal. Fingers of nanoscience/nanotechnology reach into science, engineering, computer science education, quality of life, and national security.

With nanotechnology innovations creating compounds and devices from the bottom up, there will be an impact on educational and workforce requirements. Education is also important in removing potential economic barriers for nanotechnology's use in commerce, industry, healthcare, or environmental clean-up. Just as other industries (e.g., automobile or energy) have been regulated for the national and public good, nanotechnology policy makers must approach this tsunami of scientific and technological advances with an eye for great benefits at low cost within safe margins. They must also answer ethical concerns when choosing research priorities and federally-funded applications.

Currently NNI-supported research involves the following nanotechnology societal implications:

- Help create interdisciplinary nanotechnology opportunities and exchanges
- Consider and explore the public's understanding of/outlook on nanotechnology
- Explore nanotechnology impacts on economic growth, standard of living, and competitiveness
- Involve college and university researchers in nanotechnology influences from many different areas.

With everyone having a role in how and in what direction nanoscience/nanotechnology takes, expectations and issues will be tackled as they arise. These precautions will make it possible for nanotechnology to develop in a well thought out and safe way.

International Coordination

It goes nearly without saying that something as big as nanotechnology must include the international community. By making sure that open exchange and discussion on energy, environment, health, and security are ongoing, the chance for mistakes or miscommunication decreases. Adjustments in the way international nanotechnology businesses work together will change with advances in computing, materials availability, and international trade.

Countries are forming alliances based on areas of mutual interest in nanotechnology. In 2005, Canada announced it would spend $5.5 million in research collaboration with India on the development of biotechnology and nanotechnology.

To support international communication, the National Science Foundation sponsored an international workshop in 2000 that ended in a joint declaration by participants from 25 countries and the European Union to support responsible nanotechnology development.

In January 2006, the first international symposium on nanotoxicology was held in Miami, Florida. The program included presentations and discussion on purified and non-purified nanomaterial toxicity, antioxidants, as well as tissue specific response, recognition, and clearance of nanoparticles.

Like any new venture, international nanotechnology research and development must have a plan of action that covers all bases. From economic impacts to manufacturing pitfalls and worker safety, everyone interested in nanotechnology must work together.

Bottom Line Risks and Benefits

Some people believe molecular manufacturing is possible and that the risks must be considered. They believe that it has the potential to disrupt a society for the better and or worse. (Note that most scientists dismiss the claims of these groups as science fiction.)

While it's possible that weapons and surveillance devices that are smaller and cheaper might be created, it could also be argued that future, low cost manufacturing could bring on economic upset or potential environmental damage.

Moreover, poorly planned and/or severe restrictions could lead to higher demand for risky and difficult-to-detect black-market products. In fact, all kinds of different risks can be imagined. But the truth is that any technology or practice used poorly can cause problems, and every leap in technology brings risk.

When the looming cloud of bad publicity threatens, it's important to keep in mind the great benefits nanotechnology offers:

- targeted drug delivery
- lab-on-a-chip diagnostics
- curing genetic diseases
- nanoelectronics
- nanocomposite materials
- nanosensors

- optical arrays
- fuel cells
- space exploration for industry and tourism
- efficient energy generation and transmission

Growing pains are never without discomfort. Nanotechnology has huge potential for good and evil. In fairness, the list of nanotechnology positives must be placed alongside any list of negative what-ifs. The solution is not simple. Only understanding, communication, and careful planning will prevent and avert possible dangers.

Nanotechnology is well on its way to the inexpensive production of lots of society changing products in many areas. Since the final steps of developing technologies can be much easier than the first steps, they may seem like overnight successes. The advent of widespread nanotechnology techniques may not allow enough time for everyone to adjust to its many implications.

Quiz

1. Some of the areas where nanotechnology touches society include

   (a) economic

   (b) ethical

   (c) cultural

   (d) all of the above

2. National Institute of Occupational Safety and Health plans to develop and issue a

   (a) National Nano Day proclamation

   (b) best practices document for working with nanomaterials

   (c) list of ways that everyone can have nanotubes

   (d) national Nano Prize for environmental conservation ideas

3. In one study with fish, non-functionalized fullerenes were originally found to

   (a) have an impact

   (b) be too costly to work with

   (c) be slimy

   (d) be blue in color

4. Nanotechnology has huge potential for public benefits and possible

    (a) risks

    (b) aggravation

    (c) circus acts

    (d) school lunch improvements

5. When small molecular fragments are attached to the surface of C_{60} molecules, the cytotoxic effect is

    (a) increased

    (b) decreased/eliminated, depending on the degree of functionalization

    (c) unchanged

    (d) doubled

6. Societal outlooks/impacts for nanotechnology includes all of the following except

    (a) effects on jobs

    (b) higher quality diamonds

    (c) privacy issues

    (d) medical innovations and methods

7. Which agency/interagency sponsors a wide range of research to evaluate the environmental, health, and safety impacts of nanotechnology?

    (a) INN

    (b) NNF

    (c) NNI

    (d) NTA

8. Early measurement standardization will allow scientists to compare

    (a) pay scales

    (b) pounds to meters

    (c) cashews to hazelnuts

    (d) apples to apples

9. What key component can remove potential economic barriers for nanotechnology in commerce, industry, healthcare, or environmental clean-up?

    (a) dress code

    (b) international apathy

    (c) education

    (d) laboratory space

10. With toxicity risks eliminated through functionalization, nanotechnology offers potential innovations in all of the following except

    (a) communications

    (b) water treatment

    (c) parenting

    (d) medicine

CHAPTER 14

From Here to There

It's Saturday and you're getting the house ready for a party. As you scroll through your saved settings, you try to remember which wall scene you used last time. Was it the slopes of Aspen, Colorado; undersea inhabitants off the Great Barrier Reef of Australia; or the city lights of New York? Along with your being able to tune the wall view, nanotechnology and advanced optics let you adjust the temperature and light levels to pre-programmed settings. It's a lot more convenient that the old days of wall paper and curtains.

You stop scanning for a second to look closer at the image of your youngest brother with a spoon stuck to his nose from footage taken on your birthday. You choose a view of pounding surf and sand off the coast of Hawaii—gleaming white sand and sapphire water. Since everyone at the party is supposed to wear tropical clothes, this scene should be perfect.

Now for the music. Again, you scroll through your favorite settings—jazz, rock, bongos. Volume and surround-sound add a lot to the atmosphere since you painted an entire wall with a new nano-coating that works like a single speaker. You choose the Dave Matthews Band followed by the Rolling Stones to get things started. You can change it later once everyone arrives.

Punching in the last few choices, you walk to the bedroom to get dressed, wondering how people got along without nanomaterials and all the technology/products available since the world began nanomanufacturing. Must have been like going from the horse and buggy to the jet engine!

The Big Picture

Nanotechnology, based on the principle of building with chemistry and biology one atom at a time, involves functional structures with dimensions in the 1–100 nm range. Organic chemists, in fact, have designed and fabricated nanostructures for years via chemical synthesis, but only in the past decade have developments in surface microscopy, silicon fabrication, biochemistry, physical chemistry, and computational engineering come together to provide remarkable capabilities for understanding, fabricating, and manipulating structures at the atomic level.

Nanoscience research is hot because of the intellectual appeal of crafting matter and molecules one atom at a time and because new nano methods allow the creation of materials and devices with huge public impact. The rapid evolution of nanotechnology and its applications show all the signs of becoming one of the top technologies of the twenty-first century.

With micro- and nanotechnology, science fiction is fast becoming science fact. Things like the cell phone, impossible 100 years ago without micro-electronics, are now as common as the hot dog. We can't even remember what it was like before computers, microwave ovens, and the Internet.

What exactly is going on? Are people just smarter? Probably not, Science and engineering usually builds on earlier discoveries. So when a new technology explodes as the next big thing, it's often from years of smaller "AHA!" moments filling in little knowledge pieces, until "WOW! Look what I found!"—the big picture comes into focus. Scientists and engineers are problem solvers. When they find the answer to one question, they have even more questions. When C_{60} was discovered, the idea that single atoms and molecules could be used to strengthen a material or deliver medicines in a specific spot suddenly became a reality. People got excited. The nanoscale is not just seriously small, its properties are revolutionary!

In the past 10 years, engineers have discovered that nanomaterials make metals stronger and more bendable without breaking. Pilots and aircraft engineers found that planes could be lighter with nanotube-containing metal alloys and nanocomposites and could fly farther on less fuel.

Scientists at John Hopkins University, using common metalworking techniques and nanotechnology, made a form of pure copper that is six times stronger and just as bendable as bulk copper. The copper grains, only a few hundred nanometers

across, but several hundred times smaller than bulk copper grains, help improve microelectronics and biomedical devices.

Products and Markets

Since nanotechnology crosses many scientific and engineering fields, its applications are very diverse. Nanotechnology's effects during the next decade are tough to guess because of potentially new and unanticipated applications. Table 14-1 shows an overview of the areas nanotechnology will affect in the future.

For example, if simply changing the structure of existing materials makes a big market impact, then it is possible for a whole new set of products to be created in a totally different direction from the original. However, it's likely that in the next few years most nanotechnology activity will still be in research, rather than new products.

Table 14-1 Nanotechnology improves products from the top-down and bottom-up.

Application	Bottom-Up (Nanomaterial)	Top-Down
Aerospace	Landing gear coatings, thruster and body materials	Jet engine parts, solar panels
Automotive	Corrosion-resistant coatings, spark plugs, windows, fenders	Lead acid battery electrodes
Consumer	Sports equipment, computers, hand-held devices, TVs, cosmetics, antimicrobials	Audio components/speakers
Defense	Protective armor, munitions, gun barrel coatings	Engine components
Environmental	Pollution remediation/absorption, water purification, ceramic membranes	Solar panels
Industrial Coatings	Environmental coatings (chromium, cadmium, beryllium replacements), magnetic coatings	Corrosion-resistant surface treatments
Medical	Implant coatings, antimicrobial coatings	Sensors, lab-on-a-chip
Power	Nuclear parts repair, power transformers/transmission cables	Nuclear reactor and fossil plant components, nuclear waste containment

RESEARCH TO MARKET

It is estimated in 2006 that about 460 public and private companies, 100 investors, and nearly 300 academic institutions and government entities are involved in very near-term nanotechnology applications. Many of these institutions have been able to transfer their research results into industrial applications, as shown by the rising number of patents being filed.

The number of nanotechnology patents filed at the European Patent Office in Munich, Germany, follows a similar pattern. During a nearly 20-year period (1981–1998), the number of nanotechnology patents rose from 28 to 180 patents, with an average growth rate of 7 percent in the 1990s. The United States nanotech patent growth rate was even larger from 1999 through 2003, averaging 28 percent per year, as the number of patents increased from 479 to 1011.

One important characteristic of near-term nanotechnology applications is that they have been identified as having a specific and potentially profitable use within industry and/or the consumer market. However, predicting future nanotechnology applications is sketchy. A lot of market analysts believe it's too soon to present good figures for the global market and too early to say where and when markets/applications will appear.

Some analysts believe that the peak of scientific activity is still to come, possibly by 2010, and large-scale use of nanotechnology might take until 2015.
Similarly, some forecasts have hinted at huge growth. For example, it is estimated that in 2001 approximately $50 billion went toward developing future nanotechnology markets globally. The National Science Foundation (NSF) predicted in 2000 that the total market for nanotech products and services would reach $1 trillion by 2015 and the nanotechnology analyst group Lux Research predicted in 2005 that nanotech will have a world-wide economic impact of $2.6 trillion dollars by 2015.

FIRST TO MARKET

Considering the ever-changing nanotechnology development discussion, it is important that we look at which areas of industry will be affected first. Mihail Roco, the NSF senior advisor for nanotechnology, believes that early payoffs will come in computing and pharmaceuticals, whereas others believe that medicine will be the big money market. Among the 400 nanotech products on the market today, most are nanocomposite cosmetics, coatings, textiles, sensors, and displays. They are found in everything from tennis racquets and balls to car bumpers, and from spray-on sunscreen to stain-proof trousers and skirts.

The NSF also projects that due to the high costs early on, many nanotechnology-based goods and services will be launched earlier in those markets where performance is particularly important and price secondary. Medical and aerospace

applications fit into this group. Through the experience gained, technical/ production problems would be solved and allow for the market-place introduction of new nanotechnologies.

Patents

More and more, a patents discussion is important to sort out what is new from what is a variation on the original. Nanotechnology is no exception. Companies have always used patents to protect their inventions. Universities, more than ever in the past 20 years, have awakened to the importance of getting patents on nanotechnology research processes and products.

> A *patent* is an official document open to public scrutiny that grants a right or privilege to produce, sell, or gain profit from an invention, process, or material (e.g., nanomaterial) for a set number of years (20 years in the United States).

One reason for this is that a good gauge of current and upcoming commercial nanotech activity comes from the types of patents granted. It is thought that around a quarter of all nanotechnology patents filed involve instrumentation. This points to the fact that nanotechnology is entering a development phase where the first step is to create good tools and fabrication methods.

The most progressive nanotech industrial areas are information technology (IT), chemicals, and pharmaceuticals. For example, high-capacity storage devices, flat-panel displays, and electronic paper are major IT patenting areas. Semiconductor methods such as nanoscale information processing, transmission, and/or storage devices are also important.

In chemistry and pharmaceuticals, a lot of patents are focused on finding new methods of drug delivery, medical diagnosis, and cancer or other disease treatments with huge future markets. Nanotechnology patents in other areas like construction, aerospace, and food processing show increases yearly, but their total numbers are fairly small.

Overall, it looks like IT and medicine will impact the market first as has been the pattern for the past 25 years.

Key Applications

Nanomaterials have unique, useful chemical, physical, and mechanical properties that can be used for an enormous number and variety of applications. Following are some of the big areas that will see the greatest impacts between now and 2015.

INSTRUMENTATION AND TOOLS

The tools to control nanoparticles involve different but related instruments that sense and analyze, synthesize, automate, fabricate, and service nanoscale objects and materials. Recent nanotechnology advances have come from our ability to measure and manipulate individual nanoscale structures.

Atomic force microscopes, scanning probes, nanomanipulators, optical tweezers, and other new tools allow science and engineering researchers to create new structures, measure new phenomena, and explore new applications. In the next few years, tool development will continue as a major nanotechnology catalyst. At the same time, new nanotech applications will add to the surge of specific instruments and services.

Manufacturing Tools

Cutting tools made of nanocrystalline materials such as tungsten and titanium carbide are much harder and much more wear- and erosion-resistant than their bulk cousins, so they last a lot longer. They can even machine different materials much faster, so productivity goes up. This drops manufacturing costs.

During the miniaturization of microelectronic circuits, nanocrystalline microdrill materials (drill bits with a diameter $\leq$ 100 μm) have been found to keep sharper edges and have much greater wear resistance than regular drill bits. In fact, nanocrystalline carbides are so much stronger, harder, and wear-resistant, they have improved microdrills considerably. Carbides and diamond coatings have also made an impact in large drill bits, such as those used in the oil and gas industry.

BIOENGINEERING

Some believe that nanomedicine and other nano industries will reshape the life sciences by up to 80 percent by the year 2015, as nanotechnology enables the new use of atoms, molecules, and nanoparticles within living structures/organisms.

Bioengineering is an area that combines engineering and medicine to achieve common goals. Some of the research areas include diagnostic, therapeutic, and biomolecular imaging; molecular/cellular engineering; bio-defense; and biomaterials.

> **Bioengineering** applies engineering knowledge to medicine through the use of artificial tissues, organs, and body parts (e.g., limbs, valves, pacemakers) that have been damaged, lost, or are not working correctly.

If things keep going as they are now, nanoparticles and nanodevices will open up the future of healthcare and diagnostics. The nano methods described in Chapters 5 and 6 make it possible to get past age-old diagnostic hurdles and attack disease in new ways. By accessing the nanoscale, instrument and tool advances will further medical progress.

The total nanotechnology market volume in the life sciences amounted to $8 billion in 2002. This number was expected to reach $30 billion by the end of 2006 and reach $104 billion by 2010. Similarly, the number of related companies should increase from 600 in 2002 to around 3000 by 2010.

However, it is important to remember that reported nanotechnology advances influence legal and ethical acceptance. Just as they did with genetically modified organisms (GMOs), public attitudes on life science nanotechnologies will affect nanotechnology market growth. That's why many nanotechnology researchers are working to involve the public in understanding risk and in making decisions about the ethical use of nanotechnologies.

Medical Implants

Currently, many medical implants (e.g., orthopedic implants and heart valves) are made of titanium and stainless steel alloys or special plastics such as polyethylene. These alloys and plastics are used because they are *biocompatible*. In the case of orthopedic implants (artificial bones for hip, and so on), these materials are fairly dense and non-porous.

> *Biocompatible materials* (e.g., gold alloys or plastics) do not react adversely with human tissue or cause an immune response.

For an implant to act like natural bone, the surrounding tissue has to be able to grow into the implants. This gives the implant needed strength. Since regular materials are fairly solid, human tissue can't penetrate implants well and their effectiveness decreases. Also, since plastics tend to wear out quickly (roughly 10 years), multiple, and often very expensive, surgeries are needed. This takes a toll on patients and their finances.

Bioengineering research is looking to nanomaterials for a number of uses for medical implants. For example, nanocrystalline zirconia (zirconium oxide) ceramic is hard, wear-resistant, corrosion-resistant (biological fluids are corrosive), and biocompatible. Nanoceramics can also be made porous in the form of aerogels (see Chapter 8 for more on sol-gel synthesis). These aerogels can withstand up to 100 times their weight, which translates to fewer implant replacements and a big drop in surgical costs. Nanocrystalline silicon carbide is also a good material for artificial heart valves because of its low weight, high strength, extreme hardness, biocompatibility, and wear/corrosion resistance.

Medical implants using nanoceramics will be a big benefit to everyone, especially the elderly who often suffer from osteoporosis (and take longer to heal broken bones).

COMPUTERS AND ELECTRONICS

We learned in Chapters 9 and 10 that nanotechnology offers tons of possibilities for computing, electronics, sensors, and communications. Research in nanoscience and nanotechnologies has taken off in the last decade thanks to big chunks of governmental and industrial funding. But what happens if and when a more powerful, cost-effective computing architecture comes out of future nano labs? How far will it go? Some people think that nanotechnology can create nearly unimaginable computing power that will revolutionize medicine, energy, and transportation. It's even possible that the division between computing and human consciousness will narrow quite a bit. An implantable operating system (e.g., a souped-up immune system), for example, could even be on the far horizon.

The microelectronics industry has focused on *miniaturization* for a long time. The big drive to shrink circuits and their components (e.g., transistors, resistors, and capacitors) is ongoing. When microprocessors that use these components are shrunk smaller and smaller, they become much more powerful, run much faster, and provide lightning speed computations.

A struggle in the business world for market share, products, and patents may result. Everyone would be vying for control over the latest advantages of the Information Age.

However, a lot of technical barriers still exist to computer advances including a need for ultra-fine materials in manufacturing; the ability to get rid of huge amounts of heat generated by super fast microprocessors; connectivity limitations; poor reliability; and high-quality power supplies.

Today's computers perform calculations in sequence, but quantum computers can do several things at once (in parallel). (Refer back to Chapter 10 for more on quantum computing.) For example, it would take a traditional computer billions of years to sequentially factor a 400-digit number, while a computer made of quantum particles could factor a 400-digit number in minutes. It's a completely different way of computing.

Factoring may seem like something used to test elementary school students, but it's actually needed in business and other areas. Factoring big numbers is important because it provides the basis of encryption systems that protect such things as banking transactions. If some corrupt person were to build a big enough quantum computer, businesses and even national security would be at risk. For this reason, the U.S. Defense Advanced Research Projects Agency (DARPA) and National Institute of Standards and Technology (NIST) have been funding quantum information research.

In fact, research institutions such as Oxford University; University of Innsbruck, Austria; Massachusetts Institute of Technology; California Institute of Technology; and Stanford University also have groups focused solely on quantum computing. Microsoft Research, IBM, Argonne National Laboratories and others are also probing quantum computing mysteries.

Nanocrystalline or nanoorganic materials will help the computer industry get past traditional barriers by giving manufacturers superior materials to work with. These will provide the strength, ultra-high purity, better thermal conductivity, and durable interconnections between microprocessor components needed.

Communication Antennas/Sensors

Along with smaller computer components are smaller wireless devices. Making super small, high performance antennas/sensors is an important step in the process of shrinking wireless devices and reaching a one-chip operating system. In fact, the antenna/sensor is a big off-chip component headache for computer engineers. It takes up a large amount of space (e.g., cell phone antennas) and limits just how small a device can be. To make super small, silicon-compatible antennas/sensors for advanced systems, nanotechnology and nanomaterials will be most useful with respect to their special properties and size.

OPTICS

Working with nanoscale optical information is a challenge that cuts across different scientific and engineering areas. Dielectric materials (nonconductors that direct charge, but don't stop its flow) can be used for high bandwidth communication networks and optical computing. Some applications include miniaturization of optical nanocircuits, near-field microscopy, sub-wavelength lithography and imaging, and optical sensing/monitoring of biomaterials at the cellular and molecular levels.

Look back to Chapters 5 and 6 for more biomedical applications of optic devices that produce, switch, process, and send light to different optical connections. The tough part is to make optical devices (passive and active) that can be printed on a chip for super fast and efficient system operations.

Display Devices

Television and monitor displays depend a lot on *pixel* size. Pixels are made of materials called *phosphors,* which glow when hit by an electron stream inside a cathode ray tube (CRT). The smaller the pixel/phosphor size, the better the resolution.

> A *pixel* is one of many tiny *dots* that make up the image of a picture in a display. (Pixel comes from the words *picture* + *element*).

Nanocrystalline zinc selenide, zinc sulfide, cadmium sulfide, and lead telluride made with the sol-gel technique may all lead to big improvements in monitor/display resolution. The use of nanophosphors is expected to reduce display costs and make high-definition televisions (HDTVs) and personal computers affordable for nearly everyone.

Flat-panel displays make up a huge part of the laptop computer market. Japan leads this market, mostly in research and development efforts on display materials. By making nanocrystalline phosphors, display resolution can be improved considerably and manufacturing costs cut. Additionally, flat panel displays (made from nanomaterials with enhanced electrical and material properties) have much greater brightness and contrast than conventional displays.

An *electrochromic* device is made of materials where either an optical absorption band is used, an existing band is changed with current flowing through the materials, or an electric field is applied. Nanocrystalline materials such as tungstic oxide are used in very large electrochromic display devices.

> *Electrochromism* is the ability of a material to change its optical properties when a voltage is applied across it.

The reaction controlling *electrochromism* (e.g., reversible coloration within an electric field) is the double injection of ions (or protons) and electrons that mix with the nanocrystalline tungstic acid to form tungsten bronze. These materials are used as electrochromic glazings, antistatic layers, and electrochromic layers in optical displays. Devices using this process are mostly found in public billboards and ticker boards.

> A *liquid crystal display* (LCD) is a thin, flat display device made up of any number of color or single wavelength pixels arranged in front of a light source or reflector. The pixels are made from a material that is more ordered than a liquid, but not as ordered as a crystal.

Electrochromic devices are a lot like *liquid-crystal displays* (LCD) that are commonly used in watches, calculators, and hand-held organizers. However, electrochromic devices show information by changing color when a voltage is used. When the polarity is reversed, the color is bleached (dimmed). Since the resolution, brightness, and contrast of these devices greatly depend on the gel's grain size, nanomaterials are a good candidate for use in these devices.

MATERIALS

Nanocrystalline materials created by a sol-gel technique (see Chapter 8) results in a foam-like structure called an *aerogel*. These aerogels are porous, extremely lightweight, and can hold 100 times their weight. They are made of 3D networks of particles mixed with air (fluid or another gas) trapped within their structure. Being porous and containing trapped air, aerogels are used for insulation in offices, homes, and other applications.

With aerogel insulation, heating and cooling bills are slashed, saving power and lowering the related environmental pollution. They are also good as materials for "smart" windows that darken when the sun is too bright (like self-adjusting lenses in prescription glasses) and lighten when the sun goes behind a cloud.

Ceramics

Ceramics are usually hard, brittle, and tough to machine. These qualities discourage a lot of manufacturers from using their good properties. However, with smaller grain size, ceramics are becoming more popular.

Zirconia, a hard, brittle ceramic, has even been called a *superplastic*, since it can deform to great lengths (up to 300 percent of its original length). However, ceramics must contain nanocrystalline grains to be superplastic. In fact, nanocrystalline ceramics, such as silicon nitride (Si_3N_4) and silicon carbide (SiC), have been used in such automotive applications such as high-strength springs, ball bearings, and valve lifters. They have good formability and machinability along with their excellent physical, chemical, and mechanical properties. They are also used in high-temperature furnaces.

> *Sintering* is the process of heating and compacting powdered materials at a temperature below its melting point to weld the particles into a single rigid shape.

Nanocrystalline ceramics can also be pressed and sintered into different shapes at much lower temperatures than normally needed to press and sinter regular bulk ceramics.

In fact, a range of dental appliances has been created using superplastics. In Japan, complete upper denture bases have been made using formed superplastic. To date, several thousand people have been fitted with these appliances. The dentures are said to be comfortable, lighter, and better fitting; provide better pronunciation; and prevent foods from tasting metallic or sticking to dental work.

ENVIRONMENT

As mentioned, nanocrystalline materials have really large grain boundaries (surface) compared to their grain size (volume). This makes nanomaterials' chemical, physical, and mechanical properties extremely active. Because of this high activity, nanomaterials make good catalysts that can react with toxic gases like carbon monoxide and nitrogen oxide in automobile catalytic converters and power generation equipment. In these applications, nanocrystalline materials can help prevent environmental pollution that comes from burning gasoline and coal.

High Energy Batteries

Regular and rechargeable batteries are used in almost all applications that need electrical power. Many different sized batteries are used widely in automobiles, laptop computers, electric vehicles, personal stereos, cellular phones, cordless phones, toys, and watches. Unfortunately, the life span and storage capacity of conventional and rechargeable batteries is pretty low, and they need frequent recharging.

Nanocrystalline materials made with sol-gel techniques are good choices for battery separator plates because of their aerogel structure. They hold much more energy than conventional batteries. Additionally, batteries made of nanocrystalline nickel and metal hydrides will hardly ever need recharging and will last a lot longer because of their large grain boundary (surface) area and enhanced nano properties.

Sensors

Sensors check changes in the different environments they are made to measure. These environments include changing limits of electrical resistance, chemical reactivity, magnetic permeability, heat conductivity, and capacitance. Nearly all sensors' accuracy depends on the microstructure (grain size) of the materials they are made from. A small change in the sensor's surroundings is measured by large changes in the sensor material's chemical, physical, or mechanical properties, so more and more nanomaterials are being used in detectors.

Reaction speed and extent are greatly amplified by small grain size. For example, a carbon monoxide sensor made of zirconium oxide uses its chemical stability to identify carbon monoxide. When carbon monoxide is found, oxygen atoms in

zirconium oxide react with carbon and the reaction triggers a change in the sensor's characteristics, such as conductivity. Since sensors made of nanocrystalline materials are extremely sensitive to environmental changes, they are also used in smoke detectors, aircraft ice detectors, and engine operations sensors.

AUTOMOBILES

Today's automobile engines waste a lot of gasoline. They also add to environmental pollution by incomplete gas combustion. Nanotechnology may be able to solve these problems, while increasing the automotive industry's growth and development potential. It's likely that nanotechnology and nanomaterials will affect the design/manufacturing of cars, trucks, and buses by up to 60 percent by 2015.

The ability to manipulate molecules and atoms will provide many new options for the automotive industry. For example, a regular spark plug (along with bad or worn-out spark plug electrodes) is not designed to burn gasoline completely and efficiently. Since nanomaterials are stronger, harder, and much more wear- and erosion-resistant, they are currently being considered for use in spark plugs. Nanoelectrodes would make spark plugs longer-lasting and much more fuel efficient.

An advanced new spark plug design called the *railplug* is in the prototype stages. Railplugs create powerful sparks that burn fuel better, however they also quickly corrode regular materials when used use in automobiles. However, railplugs made of nanomaterials last much longer than regular spark plugs.

Automobiles also waste big amounts of energy through engine heat loss, especially from diesel engines. Engineers are currently looking into coating engine cylinders with nanocrystalline ceramics, like zirconia and alumina, to help preserve heat efficiency and increase fuel combustion.

Better automotive brake system performance with aluminum or nanotube composites is expected to drop brake system weight (and total car weight) while increasing acceleration. Car rear windows made with carbon nanotube composites will add strength without adding weight. Plastic composite fenders, strengthened with carbon nanotubes or nanoclay, are also an important safety and performance innovation.

Overall, the automotive industry will see nanotechnology benefits from advanced power train designs, lighter weight, stronger materials, sensing technology, and higher efficiency. Since nearly all automobile components can be improved through nanotechnology, innovations and new markets are practically guaranteed, assuming that manufacturing costs can be kept down. In the next decade, automotive competition in many areas (e.g., body styles, brakes, acceleration, and safety) will depend mostly on manufacturers' ability to develop and include nanomaterials in their products.

DEFENSE

Conventional big guns, such as cannons, 155 mm howitzers, and multiple-launch rocket systems, use chemical energy from igniting chemicals (i.e., gun powder). A bullet's top speed is around 2 km/second. Electromagnetic launchers or railguns, however, use electrical energy. The related magnetic field propels projectiles (bullets) at speeds up to 10 km/second. This speed increase gives much greater kinetic energy to the same bullet mass. The higher the energy, the more target damage. For this and other reasons, the military has done a lot of research on railguns.

Additionally, since railguns run on electricity, the rails have to be good electrical conductors. They must also be strong and stiff so the railgun doesn't sag during firing and buckle under its own weight. Copper is a strong electrical conductor, but copper railguns wear out too fast. Softer copper rails, worn down by super fast, high-temperature bullets and high corrosion, makes it necessary to replace the barrel often, which raises costs.

To fix this problem, a nanocrystalline composite material of tungsten, copper, and titanium diboride is being studied. This new material has the needed strength, rigidity, hardness, electrical/thermal conductivity, and wear/erosion resistance. This results in longer-lasting, wear-resistant, and erosion-resistant railguns that can be fired faster and more often than non-nano railguns saving taxpayer money. Carbon nanotubes might also improve railgun performance even more, by providing a more durable, even higher conduction material.

Kinetic Energy Penetrators

The United States Department of Defense uses *depleted-uranium projectiles* (uranium bullets) against hardened targets and enemy armored vehicles. Although these uranium bullets have left-over radioactivity and are carcinogenic, they make use of a unique self-sharpening method on impact with a target. This, plus the lack of a non-explosive, non-hazardous replacement keeps them in use.

Nanocrystalline tungsten heavy alloys, however, have also been found to have a self-sharpening mechanism. So it's possible that nanocrystalline tungsten heavy alloys and other composites will soon replace depleted-uranium bullets, eliminating their toxic risk while keeping the sharpening, armor piercing advantage.

High-Power Magnets

A magnet's strength is related to the grain size and surface area of the material from which it is made. Magnets made from nanocrystalline yttrium/samarium/cobalt grains have strange magnetic properties because of their extremely high surface area. It's thought that applications for these high-power magnets will include quieter submarines, automobile alternators, land-based power generators, ship motors, ultra-sensitive analytical instruments, and medical magnetic resonance imaging (MRI).

Plasma TVs and Nano-Screens

Display manufacturers are building less expensive alternatives to LCD and plasma technologies. In fact, Motorola has already unveiled its first working *nano-emissive display* (NED) prototype. The wafer-thin display is just one part of a planned 42-inch wall-mounted television. It can play DVD movies as clear as on LCDs.

Instead of using cathode ray tubes or tiny LED lights to produce an image, NED uses millions of accelerated electrons charged by just 5 to 10 volts of electricity. Large-screen, high-definition LCDs use roughly 5000 volts. As the electrons speed toward a phosphor plate, a moving image is created with less power. Nano-emissive displays using carbon nanotube technology can also be seen from all angles. Additionally, NEDs are not limited by size as other display technologies. Crystal clear advertisements on huge roadside billboards and mega home entertainment units are possible.

However, as more high-definition DVD formats, television broadcasts, and cutting-edge video games grab consumers' interest, electronics manufacturers must find ways to drop production costs, make them more affordable, and get around any long term reliability problems.

LEISURE

Even leisure products have seen the coming of nanotechnology. The following areas are just a sampling of how nanomaterials have already improved current products. Table 14-2 lists the *Top Ten Nanotech Products of 2005* as named by Forbes.com.

Table 14-2 Nanotechnology looks to many different areas for future applications

Nanotechnology Product	Description	Company
1	iPod Nano – has 4GB NAND flash memory chip	Apple Computer
2	Canola Active – increases absorption of phytochemicals and lowers LDL cholesterol	NutraLease/ Shemen
3	Choco'la Chewing Gum	O'Lala Foods
4	Fullerene C-60 Day Cream - anti-oxidant action	Zelen
5	Easton Stealth CNT Baseball Bat	Easton Sports & Zyvex
6	Nanotex nano-enhanced clothing fibers- stain and water resistant	Nanotex
7	Arctic Shield - E47 polyester anti-microbial socks	ARC Outdoor
8	NanoGuard Paint – harder, water resistant, anti-mildew paint	Behr Paints
9	Activ Glass – coating on glass that cleans itself	Pilkington
10	NanoBreeze Air Purifier – oxidizing properties kills germs and pollutants	NanoTwin Technologies

Clothing

Nanotechnology has also entered the clothing industry. A windproof and waterproof ski jacket based on nanotechnology has been produced by a German company, Franz Ziener GmbH & Company by weaving nano fibers throughout the fabric. Nanotechnology makes the two-layer laminate jacket material windproof, waterproof, breathable and dirt resistant. Since it isn't an external coating, the properties won't change over time or wear away with a lot of use. Refer back to Chapter 3 for more information on stain-resistant clothing and nanofibers.

SPORTING EQUIPMENT

Golf is a game about power, accuracy, and consistency. Since its invention in Scotland in the fifteenth century, golf has been a sport that welcomes innovation. In 2004, Wilson Golf used nanometal technology to design its FwC fairway woods, Pi5 irons, and Staff Pd5/Dd5 drivers with a nanocomposite crown and a titanium shell. Golf club heads and shafts were coated with a nanometal that is stronger and lighter, and allows for greater club head speed (1 to 2 mph) and a bigger "sweet spot".

Wilson also created three high-performance PhD (Pan Head Dimple) golf balls that have dimples that are larger but nearly 50 percent shallower than standard dimples to provide greater lift and in-flight stability. Figure 14-1 shows the new Wilson products.

Figure 14-1   Wilson golf and tennis balls are new high-performance products.

Some fishing rods and tennis racquets are now using nanoparticles/nanotubes for added strength and flexibility. Other sporting equipment makers of bicycles, baseball, skis, and ice skates are looking for ways to improve performance as well. The Zyvex company in Dallas, Texas, which supplies special carbon nanotubes to several sporting goods manufacturers, is becoming a materials company in addition to being a nanotools maker.

Music

Carbon nanotubes are also being considered for additions to audio speaker cones. Increased porosity and conducting properties will give sound the fullness and feel of a concert hall. Some researchers are even working on adding carbon nanotubes to the composite materials used in special guitars and even cellos. It is hoped that they will have a bigger sound, in addition to making them tougher and lighter. Who knows, it could bring about a whole new type of nano music!

Cosmetics/Sunscreen

One of the few nanoparticle applications used in great quantity is found in cosmetics and sunscreens. Nano-enhanced cosmetics penetrate deep beneath the skin's surface layer. For example, one product uses a patented 200-nm nanotechnology process that contains vitamin A within a polymer capsule. The capsule, which acts like a sponge, soaks up and holds the cream inside until the outer shell dissolves under the skin.

The nanoparticles create a kind of surface tension that helps move the contents of the nanosome into the skin's pores. The nanosome itself dissolves on the skin's surface. This is thought to increase new cell production so that skin stays soft and free of wrinkles, even into middle age.

While companies like L'Oréal are patenting the use of dozens of different nanoparticle applications that deliver nutrients such as vitamin C into the skin, other cosmetics companies (e.g. Dior, Estée Lauder, and Johnson & Johnson) are jumping into the game with their own nanotechnology-based products.

Sunscreens are yet another nano-enhanced product. One sunscreen with SPF 30 uses Z-COTE?, a high-purity nanocrystallline zinc oxide made by BASF as its main ingredient. Usually, regular zinc oxide provides broad-spectrum protection against UVA and UVB rays, with a white chalky look that is not particularly attractive. Nano products, however, are made with nano zinc oxide and let the new sunscreen go on clear—plus, unlike cosmetics, inorganic nanoproducts aren't absorbed by the skin. They are transparent because the nanoparticles are too small to scatter light.

AEROSPACE

Aircraft builders want to make aerospace parts stronger, lighter, and longer lasting. Since strength goes down with a component's age and amount of metal fatigue, making parts from tougher materials is a high priority.

In fact, fatigue strength improves when a material's grain size drops. As we learned earlier, nanomaterials have a lot smaller grain size than conventional materials. This super small grain size provides an average gain in part life of 200 to 300 percent. Additionally, since nanomaterial parts are stronger (or lighter weight for the same strength) and some can work at higher temperatures, they allow aircraft to fly faster and more efficiently (on the same amount of fuel).

Nanomaterials are great candidates for spacecraft applications, as well. In spacecraft, high temperature resistance and material strength is critical since rocket engines, thrusters, and vectoring nozzles often work at much higher temperatures than aircraft.

Longer-Lasting Satellites

Satellites are used for both defense and commercial purposes. They use thruster rockets to maintain or change orbits as a result of factors such as the Earth's gravity and friction from the few air molecules in space, especially in low earth orbit. Satellite life is mostly set by the amount of fuel they carry. In fact, more than a third of onboard fuel is spent by partial and inefficient fuel combustion. Combustion is poor because onboard igniters wear out fast and don't perform. A nanomaterial, like nannocrystalline tungsten-titanium diboride-copper composite, offers a chance to increase igniter life and performance.

Nano-Optoelectronic Devices

Nanostructured optoelectronics offer space applications in optical satellite telecommunications and/or sensor technology (e.g., infrared sensors). *Optical wireless data links* are important for intra-satellite communication as well as optical inter-satellite links. Smaller and lighter devices having a higher bandwidth compared to common microwave communications are always needed. There is never enough bandwidth as more and more video signals are used.

Optical technology is important to data relay processing. This includes the ability to provide high data rates with low mass, low power terminals, as well as the capability to enjoy secure, interference-free communications. Earth observations, telecommunications, and science/space operations can all make use of this new method of global data transmission.

Optical links between satellites was first shown to work during the European Space Agency's ARTEMIS (Advanced Relay Technology Mission). ARTEMIS previously provided data relay for Earth observation missions that offered greater visibility and reduced image reception delays. ARTEMIS has been providing a data relay service to ENVISAT and the French national mission SPOT 4 since March 2003.

The first bidirectional optical link between the Japanese satellite, OICETS (Optical Inter-orbit Communications Engineering Test Satellite), and ARTEMIS was made on December 9, 2005. KIRARI is the second optical data relay satellite using ARTEMIS, following the earlier laser links with SPOT-4. The new satellite uses frequency-stable, solid-state lasers as data signal carriers. This optical service now operates regularly.

Space Applications for Infrared Sensors

Infrared sensors offer a variety of space applications, such as astronomy instrument sighting, satellite-based earth/atmosphere imaging research, navigation tools for space systems, and optical data communication. Improvements in infrared sensors will mostly be affected by the development of 2D (quantum well), 1D (quantum wire), or 0D (quantum dot) nanostructures.

Through quantum well or dot nanostructures, detection specifics of infrared sensors can be selectively changed to the right spectrum. This has already been done with quantum well infrared sensors, based on galadium arsenide as developed by the Center for Space Microelectronics Technology of NASA for special space applications.

Space Elevator

Beyond computers, golf clubs, and cosmetics, some nanomaterial applications (like super strong armor) are fairly far out in terms of their development time schedule. There's even talk of building a *space elevator* (i.e., a huge ladder into space that could drop off rockets and lift payloads and people into various Earth orbits). Currently, no engineering material has the strength to support a structure of such height.

However, this has not kept a lot of small startup companies, including Southwest Nanotechnologies in Norman, Oklahoma; Carbon Nanotechnologies in Houston; and Liftport Group in Bremerton, Washington, from working on this technology with NASA support with hopes of building an elevator to space by 2018.

A space elevator is not a new idea. In 1895, after visiting the Eiffel Tower, Russian scientist Konstantin Tsiolkovsky imagined constructing a "celestial castle" at the end of a spindle-shaped cable that orbited high above the clouds in a *geosynchronous orbit*. He visualized the view from such a place after watching the mechanical elevators moving slowly up and down the Eiffel Tower with packed cars of amazed passengers.

Arthur C. Clarke, well known for his book *2001: A Space Odyssey*, described a space elevator in his book *The Foundations of Paradise* that lifted satellites into orbit without rockets. Back in 1979, everyone thought Clarke's idea was imaginative but impossible, but nanotechnology has changed all that.

Before carbon nanotubes were discovered, aeronautical engineers didn't have a material strong enough to support its own weight and to tether a counterweight against the Earth's rotational (centrifugal) force. Space elevators existed only in the notebooks of inventive people. The discovery and production of standardized carbon nanotubes, however, have made a space elevator theoretically possible.

Scientists have learned out how to spin nanotubes into fiber—like wool is spun into yarn. With a light, super strong nanotube ribbon about 44,600 miles long, a space elevator, or "climber," would lift all kinds of cargo high above the Earth and release them into different orbits. The Earth's centrifugal force makes it easy for space vehicles/satellites to be "let go" to start their missions. Figure 14-2 illustrates how a climber could achieve high Earth orbit with the help of a nanotube cable and a counterweight.

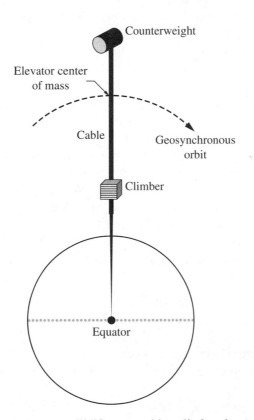

Figure 14-2   A space elevator will lift cargo with a climber that moves up and down.

Geosynchronous Orbit

When an orbiting object stays at the same place over the spinning Earth, it is in a *geosynchronous orbit* (i.e., geostationary). Geostationary satellites are positioned over points on the equator. The Earth's gravity and the centripetal (center-seeking) force of the orbiting object keeps it in the same spot over the Earth without it getting ahead or falling behind. To be in a geosynchronous orbit, an object (e.g., satellite) must be at 35,900 km (22,300 miles) above the Earth's surface and rotate at the same speed 1670 km/hr (1,038 mph) as the Earth rotates.

Why build an elevator to space? Two reasons: a better life on Earth and cheaper space travel. Today, it costs around $10,000 per pound to send the NASA Space Shuttle into a low earth orbit. It costs roughly $20,000 per pound to send a satellite into geosynchronous orbit. A space elevator would cut that cost to around $400 per pound or less. Communications would become cheaper and more available. Energy would get cheaper, too, since huge solar collectors could be lifted into space to beam back clean, unlimited energy to ground stations. Pharmaceuticals and crystals for electronics would also be easier to make (purer and cheaper) without gravity's pull. It would also be much cheaper to launch space probes to explore the solar system or build an exploratory moon base, not to mention fuel space tourism in a big way.

Can it be done?

The best place for a space elevator cable to be anchored would be a movable platform located at the Earth's equator in the Pacific Ocean. The absence of trade winds along the equator's corridor makes the area a lot safer as a potential elevator site, since hurricane threats are nearly non-existent.

Additionally, since the Pacific Ocean is so vast, a space elevator platform could be placed outside of normal shipping lanes. Air space around the cable would also be restricted. This would be monitored and protected as other national restricted airspace is protected.

However, before a space elevator can be built, a few technical and security concerns must be considered:

- Orbital junk like old, dead satellites
- Meteorites and high speed micrometeorites
- Corrosion from atmospheric oxygen
- Weather problems like lightning and wind
- Fixing ribbon breaks near the ground, half-way, or at the far end
- Wild vibrations
- Radiation shielding
- Sabotage

The companies interested in space exploration and a space elevator have included these potential snags in their design and operational plans. Who knows, some day in the future, we may look up weather forecasts for a ride on a space elevator as commonly as we do for other travel destinations.

Nano Worldwide

When considering the synthesis and processing of nanostructures from now until roughly 2020, it's plain to see that organic, inorganic, and biological applications are underway. The major forces will be creativity, opportunities, applications, and economics in the areas of science, medicine, and technology. Greater improvements in synthesis and assembly, along with a high degree of precision, will be gained through innovative processing.

A combination of top-down assembly methods with bottom-up chemical and biological assembly techniques will be needed to make fully functional nanostructures operational at the nanoscale. Once this is reliably accomplished, the size, shape, structure, and characteristics of molecular connectivity, nano-devices, and nanostructured materials will be established along with the next generation of many, many products.

By 2015, world markets for pure nanotech products will expand into the billions in many industries. Table 14-3 describes the near-term and far out applications predicted with nanoscience and nanotechnology.

Worldwide, more than 4000 companies and research institutes are working with nanotech, including about 1900 companies in services and 1020 companies making products. Total worldwide nanotechnology markets are predicted to grow from $300 billion in 2006 to more than a trillion dollars in 2015.

Currently, the leading countries are USA, Japan, China, and Germany. China is now one of the world's leaders in terms of newly established nanotechnology firms. From 2003 to 2006, the number of nanotechnology companies in China has grown to more than 600. This huge growth rate is still rising.

Overall, surging global interest has highlighted nanomaterials and nanostructures as important in improving current products. And, as more is known about their amazing properties, nanotechnology applications will be seen as profitable future investments. Far off revolutionary advances in nanotechnology can only be dreamed of today. In fact, a lot look like science fiction, but it's possible that the nanotechnology dreams of tomorrow will be reality some day. In nanotechnology, great things come in super small packages.

Table 14-3 Nanotechnology applications projected for the future

Product/Application	Description
Full-wall video screens	Television and video game enhancement
Full-wall speakers	Television, video games, sound systems enhancement
Programmable paint	Changes color, texture, and pattern on command
Reprogrammable books	Look and feel like a book, but can change content
Reusable paper/canvas	Drawings can be changed, stored, or done in different media (e.g., oils, watercolors, etc.)
Self-adjusting contour chairs	Change shape to fit the person sitting in them
Paint-on board games	Applied to any surface; finds edges and scales to space available
Board games with millions of parts	Allows economic, logistical, and military simulations
Variably transparent windows/ walls	Changes amount of transparency or opaqueness related to set interior light level and outside light level changes
Retractable walls and ceilings	Rooms can be changed to meet space and use needs
Walk-through walls	With million of spring-load hinges that allow it to break apart when you walk into it and snap back together after
Programmable rooms	Configurable walls, ceilings, floors, and optics (e.g., two-room house with the other room reconfigurable to whatever you need)
Always sharp knives	Nanoparticles combined that never become dull
Reversible seals	Seal and unseal on command
Always clean, non-slip bathtubs	Dirt, minerals don't stick
People scrubbers	Run off of body heat to remove dirt, oils, odors, so you never need a shower
Spray-on clothes	Pants, shirts, foul weather gear
Temperature-sensitive cloth	Weave density changes according to outside temperature
Paint-on watches and thermometers	Instant information; no more lost watches
Temperature-sensitive gloves	Receptors convey pressure and texture but not heat and cold
Time-release medicine	Site-specific and dose-dependent
Whole-building shock absorbers	For earthquake and tornado areas
Self-forming scale models	From computer assisted drawing designs
Changeable molds and forms	Building materials from concrete and plastics
Paint-on white boards	Can be read by a computer when used with a special pen

Table 14-3   (continued)

Product/Application	Description
Self-testing construction material	Constantly checks structural integrity and current load, beeps when unsafe or beyond tolerance limits
Force-sensitive tires	Tires change shape to meet changing road and driving conditions
Surrounding bumpers	Extend when object is too close then change back afterward
Computer-generated holograms	Use easily available computer power
Contact lens virtual reality	For the complete gaming experience
Full surround virtual reality	Where experience completely surrounds the user with visual, auditory, tactile, and realistic physical feedback
Space elevator	With ribbon-like cables that extend beyond the atmosphere
Cellular diagnosis	Nano devices that identify and repair cell and protein damage
Implantable computers	Hard drive is always with you
Implantable translators	No more language barriers
Anti-wrinkle moisturizer	No more wrinkles
Time release perfumes	Stays fragrant on skin for a week
Cosmetics	Reduce irritation/allergic reactions; promotes skin cell health
Space elevator	Super strong thin cable
Sports equipment	Equipment, analyzes areas for improvements
Anti-fouling/anti-corrosion coatings	Improved boat/vehicles life, engines, machinery

Quiz

1. The National Science Foundation predicts that the total market for nanotech products and services will reach $1 trillion by

    (a) 2007

    (b) 2010

    (c) 2012

    (d) 2015

2. An official document that grants a right to produce, sell, or get profit from an invention, process, or material (e.g., nanomaterial) for a set number of years is called a

    (a) hall pass

    (b) patent

    (c) menu

    (d) will

3. During a nearly 20-year period, the number of nanotechnology patents in Europe rose from 28 to

    (a) 80 patents

    (b) 100 patents

    (c) 180 patents

    (d) 200 patents

4. When golf club heads/shafts are coated with a nanometal they become

    (a) slippery

    (b) stronger and lighter, with greater club head speed

    (c) golden

    (d) slower, heavier and more dense

5. As of 2006, about how many companies and research institutes are working with nanotechnology?

    (a) 1200

    (b) 2850

    (c) 3500

    (d) 4000

6. The process of heating and compacting powdered materials at a temperature below its melting point to weld the particles into a single rigid shape is called

   (a) compaction

   (b) smelting

   (c) sintering

   (d) smoldering

7. A combination of top-down assembly methods with bottom-up chemical and biological assembly techniques will be needed to make

   (a) fully functional nanostructures operational at the nanoscale

   (b) surround-sound audio systems

   (c) a new line of swimming suits

   (d) nanoaluminum powders

8. The best place for a space elevator cable to be anchored would be a movable platform located at

   (a) the North Pole

   (b) Times Square, New York

   (c) Mount Everest

   (d) the Earth's equator

9. Which field applies engineering knowledge to medicine through the use of artificial tissues, organs, and body parts that have been damaged, lost, or are not working right?

   (a) bioengineering

   (b) archeology

   (c) geology

   (d) nuclear engineering

10. While a traditional computer would take billions of years to sequentially factor a 400-digit number, a computer of quantum particles could factor it in

    (a) seconds

    (b) minutes

    (c) hours

    (d) days

Part Four Test

1. What organization plans to create an information bank to keep everyone current with the latest nanoscience/nanotechnology discoveries and public policies?

   (a) Environmental Protection Agency

   (b) National Nanoscience Council

   (c) International Council on Nanotechnology

   (d) Institute of Higher Learning

2. What is a marketing firm's dream topic to advertise right now?

   (a) cryptology

   (b) nanotechnology

   (c) athletic shoes

   (d) oceanography

3. The strongest molecules known before C_{60} were

   (a) lead

   (b) gold

   (c) krypton

   (d) diamond

4. VivaGel is a

   (a) hair care product

   (b) lubricant

   (c) lip gloss

   (d) microbicide

5. Who were among the first nanotech visionaries to dig down to the nanoscale world?

   (a) chip manufacturers

   (b) archeologists

   (c) optometrists

   (d) building contractors

6. The official standard by which all other standards are compared is made from what material?

   (a) platinum

   (b) iron

   (c) silver

   (d) tungsten

7. The toxicity of nano-C_{60} and water-soluble fullerenes is connected to their level of

   (a) speciation

   (b) derivatization

   (c) acidification

   (d) distribution

8. For how many more years will silicon chips keep getting better/smaller?

   (a) 1–4

   (b) 5–10

   (c) 11–20

   (d) 21–40

9. Theoretically, assemblers are able to

(a) take over the world

(b) put bicycles together from the directions

(c) pull atoms from molecules to build other molecules

(d) make uncoordinated people better dancers

10. Therafuse Inc. in Vista, California, is creating a skin patch for

(a) rabies

(b) diabetes

(c) avion flu

(d) bed wetting

11. Nanowire technology was developed by Harvard University chemist

(a) Charles M. Lieber

(b) Denise Williams

(c) Elisabeth Caradec

(d) Larry Bock

12. An environmental concern regarding the use of nano-iron for toxic cleanup is

(a) distribution

(b) recovery

(c) rust

(d) demarcation

13. Nanotechnology products have at least one dimension in the following range.

(a) $\geq$ 100 millimeter range

(b) $\geq$ 10 millimeter range

(c) $\leq$ 100 nanometer range

(d) $\leq$ 1 decimeter range

14. When an object travels at the same speed as the spinning Earth, it is in a

(a) acute orbit

(b) galactic orbit

(c) solar orbit

(d) geosynchronous orbit

15. What common interest does IBM, Hewlett-Packard, DuPont, General Electric, Motorola, Sony, Siemens, and Xerox share?

    (a) development of chemical solutions

    (b) marginal stock prices

    (c) nanotechnology

    (d) low dollar management

16. Regulators who have to identify and define nanomaterials need standard

    (a) business attire

    (b) crystalline structures

    (c) forms

    (d) nomenclature

17. According to NanoMat, a materials research group, products with some form of nanoparticle/nanomaterial components were worth how much in 2004?

    (a) $12.2 billion

    (b) $22.4 billion

    (c) $26.5 billion

    (d) $28.4 billion

18. Great conductors, nanotubes may be able to keep electric car batteries charged by reclaiming lost

    (a) mileage

    (b) heat energy

    (c) time

    (d) gasoline fumes

19. Of the following governmental agencies, which is not working with the National Nanotechnology Initiative?

    (a) National Science Foundation

    (b) National Institute of Environmental Health Sciences

    (c) Department of Energy

    (d) Department of Current Affairs

20. Factoring big numbers is important because it provides the basis for what kind of systems that protect banking transactions?

    (a) monster

    (b) electrical

    (c) entertainment

    (d) encryption

21. A necessary test of a sample's properties through carefully recorded observations and measurements is a(n)

    (a) law

    (b) experiment

    (c) mistake

    (d) hybridization

22. Job impacts, medical advances, environmental effects, and privacy issues from nanotechnology development all affect

    (a) seasonal temperatures

    (b) academic status

    (c) society

    (d) test scores

23. In 2004, Wilson Golf used nanometal technology to design its FwC fairway woods, Pi5 irons, and Staff Pd5/Dd5 drivers with a

    (a) weekend golf pro

    (b) tin shaft and iron head

    (c) more leisurely swing in mind

    (d) nanocomposite crown and a titanium shell

24. Nanostructures with branching physical characteristics that make them great tools to target diseases and deliver drugs are called

    (a) dendrimers

    (b) muckyballs

    (c) detritus

    (d) anolimbs

25. When compound A is changed into product B with a slightly different reactivity, solubility, and/or chemical composition, it is called a

    (a) nucleic acid

    (b) valence electron

    (c) derivative

    (d) isotherm

26. The National Science Foundation projects nanotechnology as a $1 trillion market by

    (a) 2010

    (b) 2015

    (c) 2020

    (d) 2025

27. What fraction of onboard satellite fuel is spent by partial and inefficient fuel combustion of repositioning thrusters?

    (a) 1/4

    (b) 1/3

    (c) 1/2

    (d) 2/3

28. Air space around the space elevator cable would be

    (a) a tourist attraction

    (b) cloudy

    (c) restricted

    (d) unrestricted

29. Using a computer programmed to study anthrax spore surfaces, NanoInk, Inc., will produce

    (a) glow in the dark ink

    (b) super thin fountain pens

    (c) detection kits

    (d) food colorings

30. An important angle of nanotechnology risk concerns nanoparticles' extreme

    (a) mobility

    (b) odor

    (c) non-conductance

    (d) brittleness

31. Companies like L'Oréal are patenting the use of dozens of different nanoparticle applications that deliver nutrients into the skin such as

    (a) vitamin B_{12} in olive oil

    (b) vitamin C

    (c) botox

    (d) nanophosphors

32. To most of the public, nanotechnology seems like

    (a) yesterday's news

    (b) engineering lingo

    (c) science fiction

    (d) a buzz word

33. Optical links between satellites was first shown to work during which European Space Agency mission?

    (a) ARTEMIS

    (b) ARROW

    (c) CALLISTO

    (d) ORION

34. Toxic fullerenes affect cell membranes by the production of

    (a) lipids

    (b) amino acids

    (c) nitrogen

    (d) oxygen radicals

35. Oxford University; University of Innsbruck, Austria; Massachusetts Institute of Technology; California Institute of Technology; and Stanford University all have groups focused on

    (a) ice skating

    (b) quantum computing

    (c) climbing the Himalayas

    (d) building a better mouse trap

36. Growing investments in nanotechnology have been compared to the

    (a) dot com bubble

    (b) story of David and Goliath

    (c) penny saved is a penny earned slogan

    (d) California gold rush days

37. All of the following are important concerns of building a space elevator except

    (a) a company interested in designing the project

    (b) radiation shielding

    (c) vibrations

    (d) sabotage

38. One way for the government to "get it right" with nanotechnology, is to promote a

    (a) lot of big spending

    (b) sudden stop to all research

    (c) gradual introduction

    (d) special tax on biochips

39. Using nanotubes, bionanotechnology will make biochip chemical tests

    (a) 10 times more sensitive

    (b) 1000 times more sensitive

    (c) 10,000 times more sensitive

    (d) 100,000 times more sensitive

40. Z-COTE, a high-purity nanocrystalline zinc oxide that protects against UVA and UVB rays is made by

    (a) BASF

    (b) ACME

    (c) NASA

    (d) INTEL

Final Exam

1. All of the following are DNA nitrogenous bases except
   (a) guanine
   (b) cytosine
   (c) adenine
   (d) tyrosine

2. Tests at the National Institutes of Environmental and Health Sciences and the Federal Drug Administration are looking at what kind of exposure route for engineered nanoparticles?
   (a) skin
   (b) synthetic
   (c) geothermal
   (d) tattoos

3. Depending on which medicine molecules are encapsulated, doctors would be able to change the

    (a) cost of medicine

    (b) necessity of taking medicine

    (c) strength and type of medicine prescribed

    (d) office staff's workload

4. Killer organisms, doomsday weapons, and science run amok are science fiction, not

    (a) film themes

    (b) nanotechnology

    (c) archaeology

    (d) inappropriate for young children

5. Attaching small molecular fragments to the surface of C_{60} molecules is called

    (a) shear stress

    (b) functionalization

    (c) characterization

    (d) optimization

6. Proteins are

    (a) found within inorganic metals

    (b) known to be colorless compounds

    (c) made of long chains of amino acids that fold into specific structures

    (d) composed of wood and minerals

7. A particle is considered to be in the nanoscale when one of its dimensions is less than

    (a) 1 meter in length

    (b) 50 centimeters in length

    (c) 100 millimeters in length

    (d) 100 nanometers in length

8. Nanoparticles' special properties are primarily based on their

   (a) size

   (b) temperature

   (c) solidity

   (d) smell

9. As far as importance to humans as a tool, nanotechnology is the hottest thing since

   (a) fire

   (b) water

   (c) wind

   (d) trees

10. When more than 1000 single cell assays are combined on a chip the size of a thumbnail, it is known as

   (a) potato chip

   (b) fish and chips

   (c) chocolate chip

   (d) lab-on-a-chip

11. In the United States, approximately how many untreated Superfund sites exist?

   (a) 100

   (b) 500

   (c) 1000

   (d) 10,000

12. The symbol on the periodic Table for silver is

   (a) Si

   (b) Ag

   (c) Hg

   (d) Sr

13. Nanotechnology creates materials from

    (a) atoms and molecules

    (b) water

    (c) soil

    (d) cosmic rays

14. In 1870, Lothar Meyer's next generation of the Periodic Table contained

    (a) 28 elements

    (b) 57 elements

    (c) 74 elements

    (d) 112 elements

15. Before nanoscale imaging instruments became available, scientists and physicians depended on all of the following diagnostic methods except

    (a) astrology

    (b) X-ray

    (c) ultrasound

    (d) tissue biopsies

16. The EPA has lowered the arsenic standard for drinking water to

    (a) 1 part per billion

    (b) 5 parts per billion

    (c) 10 parts per billion

    (d) 50 parts per billion

17. When three tightly wound protein strands combine into long, thin molecules, they form the abundant protein molecule called

    (a) hemoglobin

    (b) collagen

    (c) albumin

    (d) human growth factor

18. Which negatively charged subatomic particles orbit around an atom's positively charged nucleus?

    (a) neutrinos

    (b) protons

    (c) electrons

    (d) quarks

19. Professor Richard Feynman was a

    (a) physicist

    (b) biologist

    (c) sociologist

    (d) pathologist

20. The main sources of atmospheric nitrogen oxide include all of the following except

    (a) lightning

    (b) mountain streams

    (c) forest fires

    (d) volcanoes

21. Nanotechnology is hugely

    (a) locomotive

    (b) multi-disciplinary

    (c) hyperactive

    (d) fantasy

22. What's the biggest problem with solar power?

    (a) too reliable

    (b) not enough watts from the sun each day

    (c) creates too much pollution

    (d) night time

23. The physics of DNA and RNA movement through nanochannels has a direct effect on future

    (a) family planning

    (b) filtration methods and membrane design

    (c) hair and eye color

    (d) environmental remediation of Technetium-7

24. What percent of scientists polled admit that current atmospheric CO_2 concentrations are going to cause severe problems over the next 100 years?

    (a) 24%

    (b) 45%

    (c) 72%

    (d) 99%

25. With toxic chemotherapy drugs, there is often a fine line between killing cancer cells and

    (a) causing rashes

    (b) patient stress levels

    (c) killing the patient

    (d) tackling medical insurance

26. Alchemy was the early name for

    (a) chemistry

    (b) botany

    (c) computer science

    (d) mathematics

27. The number of nano-related papers increased from 0 in 1990 to how many in 2005?

    (a) 200

    (b) 4,032

    (c) 20,000

    (d) 32,000

28. A semiconductor nanocrystal of a few nanometers to a few hundred nanometers in overall size is called a

    (a) quantum dog

    (b) quantum dot

    (c) bacteria

    (d) amoeba

29. Don Eigler, who wrote the letters *IBM* with xenon atoms, later developed an

    (a) ice cream machine that takes orders

    (b) anti-glare laboratory fume hood

    (c) anti-fouling coating for navy ships

    (d) electrical switch that turns on and off using an atom

30. Nanotechnologists primarily use the following tool types except

    (a) inspection tools

    (b) manufacturing tools

    (c) culinary tools

    (d) modeling tools

31. Water quality is affected by all of the following except

    (a) extracted and burned fossil fuels

    (b) deforested and developed land

    (c) manufacturing process by-products

    (d) smooth river rocks

32. Whose work led others to believe that the majority of an atom's mass is in its nucleus?

    (a) Ernest Rutherford

    (b) J.J. Thomson

    (c) Hans Geiger

    (d) Ernest Hemingway

33. When scientists perform tests outside a living organism, it is known as

    (a) *in centro*

    (b) *in vivo*

    (c) *in livo*

    (d) *in vitro*

34. When a substance added to a liquid increases its spreading or wetting properties by reducing its surface tension, it's known as

    (a) an oil slick

    (b) a surfactant

    (c) molasses

    (d) a big mess

35. When a material is said to be cytotoxic, it means that it

    (a) kills cells

    (b) grows well

    (c) is found in granite

    (d) is a pale blue/green

36. What Russian scientist imagined constructing a "celestial castle" orbiting high above the clouds at the end of a spindle-shaped cable?

    (a) Steve Smolen

    (b) Konstantin Tsiolkovsky

    (c) Paul E.F. Caradec

    (d) Mikhail Baryshnikov

37. In order to predict the speed and efficiency of nanoparticles' movement through water and soil, it's critical to understand

    (a) optical properties

    (b) transport methods

    (c) water salinity

    (d) remediation costs

38. Once heavy metals such as lead and mercury are oxidized by iron they become insoluble and

   (a) green

   (b) radioactive

   (c) fertilizer

   (d) locked within the soil

39. Today the Earth supports approximately how many people?

   (a) 3 billion

   (b) 4 billion

   (c) 5 billion

   (d) 6.5 billion

40. When a lot of short DNA molecules are bound to a solid surface, it is known as a DNA or

   (a) chocolate chip

   (b) silicon chip

   (c) biochip

   (d) neochip

41. The first bi-directional optical link between ARTEMIS and which satellite was made on December 9, 2005?

   (a) Russian

   (b) Japanese

   (c) Chinese

   (d) Indian

42. Which nuclear particles have no charge?

   (a) electrons

   (b) protons

   (c) neutrons

   (d) sienna soot

43. Magnification is the amount that an image is enlarged under a

    (a) stethoscope

    (b) microscope

    (c) gyroscope

    (d) altimeter

44. Total nanotechnology worldwide markets are projected to grow from $300 billion in 2006 to

    (a) nearly $385 billion in 2015

    (b) nearly $500 billion in 2015

    (c) nearly $700 billion in 2015

    (d) nearly $900 billion in 2015

45. Who was the first person to see live bacteria?

    (a) Alexander Fleming

    (b) Leonardo di Vinci

    (c) Antony van Leeuwenhoek

    (d) John Bertouli

46. An object in a geosynchronous Earth orbit is

    (a) an impossibility

    (b) in different spots over the Earth

    (c) always tethered at the North Pole

    (d) in the same spot over the Earth

47. What type of imaging is used to make in *vivo* molecular actions visible, quantifiable, and traceable over time in humans or animals?

    (a) mutagenic imaging

    (b) molecular imaging

    (c) meteorological imaging

    (d) comic book imaging

48. What sports equipment manufacturer created a line of golf drivers that included nanomaterials?

    (a) Head

    (b) Nike

    (c) Wilson

    (d) Everlast

49. For researchers to discover and investigate new chemical, physical, and biological happenings in the nano world, they need

    (a) high-tech tools

    (b) lots of caffeine

    (c) more vacation days

    (d) Fridays off

50. The ability of a material to change its optical properties when a voltage is applied across it is known as

    (a) high market value

    (b) electrochromism

    (c) phototropism

    (d) radioactivity

51. When light is shown on this nanostructure, a strong electromagnetic vibrating field (in the near infrared) is created in and around it. Which nanostructure?

    (a) carbon nanotube

    (b) gold wedding ring

    (c) nano hydrogen

    (d) gold nanoring

52. SWNT stands for

    (a) sweet and natural turnips

    (b) seldom-witnessed nocturnal tap dancing

    (c) single-walled carbon nanotubes

    (d) stationary wired nanomaterial testing

53. Nanotechnology is to energy and medicine what plastic was to

    (a) take-out containers

    (b) cosmetics

    (c) string theory

    (d) beach sand

54. Infrared sensors offer all the following space applications except

    (a) space system navigation tools

    (b) self-sharpening probes

    (c) optical data communication

    (d) satellite-based earth/atmosphere research

55. A nanoparticle has one or more dimensions of

    (a) 1–100 nm

    (b) 100–1000 nm

    (c) 10,000–100,000 nm

    (d) 100,000–1 billion nm

56. The National Nanotechnology Initiative provides coordination and direction for what new field?

    (a) biotechnology

    (b) agronomy

    (c) biochemistry

    (d) nanotechnology

57. Single quantum dots can be formed by a method called

    (a) archeology

    (b) lithotripsy

    (c) electron beam lithography

    (d) candy making

58. A de Broglie wavelength is the measure of wave movement (wavelength) of a

    (a) tidal wave

    (b) flag

    (c) meteor

    (d) particle

59. Carbon nanotubes, predicted to have unique electrical and mechanical properties, have now demonstrated these properties through

   (a) individual nanotube measurements

   (b) high funding levels

   (c) low pressure marketing

   (d) luck and persistence

60. In the soil, iron particles are not changed by levels of

   (a) soil acidity

   (b) temperature

   (c) nutrients

   (d) all of the above

61. Nanotechnology can best be called

   (a) too good to be true

   (b) better than bulk

   (c) science/engineering worth watching

   (d) mind over machine

62. Electron beam lithography is used to form

   (a) top-of-the-line pens

   (b) high end radios

   (c) single quantum dots

   (d) aerogels

63. A material composed of two or more metals (e.g., brass is made from copper and zinc) or a metal and a non-metal is known as

   (a) a soufflé

   (b) an alloy

   (c) an ideal gas

   (d) compost

64. When two or more phases are present (one dispersed into the other) in a combined material, it is known as

    (a) an omelet

    (b) a colloid

    (c) a science fair project

    (d) an inert material

65. The main advantage of diagnostic saliva tests is that they are

    (a) non-sticky

    (b) invasive

    (c) renewable

    (d) non-invasive

66. EcoTru Professional stamped out what percentage of post-operative infections when used in Africa?

    (a) 25%

    (b) 50%

    (c) 75%

    (d) 100%

67. Unlike electronic devices that use electron charge to carry signals, quantum computing uses

    (a) heat

    (b) electron spin or light polarization

    (c) wind energy

    (d) quarks

68. A semiconductor nanocrystal (a few nanometers to a few hundred nanometers in overall size) is known as a

    (a) quantum rot

    (b) quantum spud

    (c) quantum dot

    (d) quantum tot

69. Antony van Leeuwenhoek created a simple microscope that magnified samples up to how many times?

    (a) 50

    (b) 75

    (c) 100

    (d) 200

70. What color do blocked X-rays appear on X-ray film, providing contrast to healthy tissue?

    (a) white

    (b) black

    (c) purple

    (d) brown

71. Bioengineers at the University of California, San Diego, are working with nanocrystals that can be instructed to

    (a) absorb excess carbon monoxide from the lungs

    (b) play the top ten music hits of the week

    (c) shine only at night or in months that contain an *r*

    (d) move toward tumor blood or lymphatic vessels

72. Arthur C. Clarke imagined lifting satellites into orbit without rockets in his book *The Foundations of Paradise in*

    (a) 1865

    (b) 1979

    (c) 1986

    (d) 2002

73. What type of nanotechnology methods provide direct molecular measurements of folding forces and dynamics?

    (a) brute strength

    (b) tiny vises and clamps

    (c) optical tweezers

    (d) super small pliers

74. TEM can see images how many times smaller than a compound microscope?

    (a) 500

    (b) 1000

    (c) 25,000

    (d) 100,000

75. Who pieced together DNA's structure in 1951?

    (a) Watson and Crick

    (b) James and Abby

    (c) Williams and Bennett

    (d) Bryn and McKenna

76. The ability of a lens or optical system to form separate and distinct images of two objects with small angular separation is known as its

    (a) availability

    (b) resolving power

    (c) contactivity

    (d) defensive power

77. The major toxicological effects of engineered nanoparticles on living organisms are

    (a) well defined

    (b) still a mystery

    (c) partially understood

    (d) not important

78. One of nanocrystals' biggest advantages over larger materials is that their size and surface can be

    (a) easily cracked

    (b) used as a facial mirror

    (c) hidden from prying eyes

    (d) precisely controlled and properties tuned

79. Transmission electron microscopes look through a sample like a

    (a) slide projector

    (b) magnetic mirror

    (c) rhesus factor

    (d) hypotenuse

80. Like earlier times (Iron Age, Bronze Age, Industrial Age, and Information Age), we are now entering the

    (a) Polluted Age

    (b) Euro Age

    (c) Molecular Age

    (d) Celebrity Age

81. Tunneling current is created when a surface dip, hole, or groove causes the voltage to dip as the electrons between the probe tip and sample

    (a) move together

    (b) get farther apart

    (c) stay the same

    (d) no current is used to examine sample surfaces

82. CBEN is an acronym for the

    (a) Center for Bananas, Eggplant and Nectarines

    (b) Center for Biological and Environmental Nanotechnology

    (c) Consolidated Biostatistics for European Nanotechnology

    (d) Council on Basic Environmental Nanoscience

83. Aerosols and foams are types of

    (a) colloids

    (b) nano hair gel

    (c) metamorphic rock

    (d) DNA

84. Fishing rod and tennis racket manufacturers have now included nanoparticles/nanotubes in their products for added

    (a) hype

    (b) sales

    (c) strength and flexibility

    (d) color

85. The United States and other governments have spent billions of dollars on chemical runoff and toxic landfill clean up because

    (a) everyone knew of the toxic dangers

    (b) they didn't have anything else to do with the money

    (c) "new and improved" was considered always good

    (d) regulations and standards were strictly enforced initially

86. The higher the number of molecular fragments attached to the surface of C_{60} molecules, the greater the reduction of

    (a) cost

    (b) expressivity

    (c) substrate variety

    (d) cytotoxicity

87. A surfactant is a substance (e.g., detergent) added to a liquid that increases its spreading properties by reducing its

    (a) odor

    (b) melting point

    (c) surface tension

    (d) cost

88. Carbon nanotubes have super strength as much as

    (a) 20 × the strength of steel

    (b) 50 × the strength of steel

    (c) 70 × the strength of steel

    (d) 100 × the strength of steel

89. A type of electric motor based upon a material's change in shape in an electric field is called a

    (a) outboard motor

    (b) piezoelectric motor

    (c) sequestration motor

    (d) retrorocket

90. Experiments with genetic protein molecules like DNA and RNA show movement through membrane channels via

    (a) electrical currents

    (b) salt

    (c) collagen

    (d) self-assembly

91. Which nanomaterials will most likely make a working space elevator possible?

    (a) nanocrystals

    (b) nanotubes

    (c) buckyballs

    (d) nanopowders

92. The nanoscale is not just another rung on the miniaturization ladder, but is

    (a) science fiction

    (b) antiquated

    (c) only for people trying to lose weight

    (d) a qualitatively new scale

93. Which of the following is most used as a radioactive tracer to detect heart disease?

    (a) cobalt

    (b) sulfur

    (c) thallium

    (d) rhodinium

94. In the areas of theory, modeling, and simulation, most nanotechnology progress has been connected with the introduction of more powerful

    (a) computers

    (b) acids

    (c) politicians

    (d) bases

95. Buckyballs, single-walled nanotubes, nanoshells, quantum dots, and microcapsules have been called

    (a) funding buzz words

    (b) impossible to create

    (c) too complex to fool with

    (d) smart materials

96. Professor Richard Smalley stated that new nanomaterials were key to solving which of humanity's problems?

    (a) energy

    (b) funding

    (c) population

    (d) environment

97. Nanotechnology is also called

    (a) molecular mud

    (b) molecular hypnosis

    (c) molecular manufacturing

    (d) molecular fiction

98. In nations such as Bangladesh, India, Mexico, Argentina; Taiwan, and Thailand, what percentage of the population is afflicted with arsenic poisoning?

    (a) 1–5%

    (b) 10–40%

    (c) 50–75%

    (d) 90–95%

99. A laser scanning, confocal microscope uses a laser light (ultraviolet) and scanning mirrors to sweep across what kind of a sample?

 (a) ice cream

 (b) fluorescent

 (c) atmospheric

 (d) liquefied petroleum gas

100. EcoTru is a non-toxic, yet potent

 (a) disinfectant

 (b) tree trimmer

 (c) fertilizer

 (d) cotton-based clothing line

Answers to Quiz, Test, and Exam Questions

CHAPTER 1

  1. D   2. C   3. D   4. B   5. D

  6. C   7. B   8. C   9. B   10. A

CHAPTER 2

  1. D   2. C   3. B   4. B   5. C

  6. D   7. A   8. A   9. C   10. D

CHAPTER 3

1. B   2. D   3. C   4. B   5. B

6. A   7. C   8. D   9. C   10. A

CHAPTER 4

1. A   2. B   3. B   4. C   5. D

6. D   7. C   8. D   9. D   10. B

PART ONE TEST

1. C   2. A   3. D   4. B   5. B

6. D   7. B   8. C   9. D   10. A

11. C   12. B   13. A   14. B   15. B

16. A   17. C   18. C   19. B   20. D

21. B   22. B   23. A   24. C   25. D

26. A   27. C   28. D   29. B   30. C

31. C   32. B   33. C   34. D   35. A

36. C   37. C   38. B   39. B   40. D

CHAPTER 5

1. C   2. D   3. A   4. C   5. B

6. C   7. A   8. B   9. D   10. B

CHAPTER 6

1. C   2. B   3. D   4. A   5. C

6. B   7. C   8. A   9. D   10. B

CHAPTER 7

1. B   2. A   3. C   4. C   5. B

6. D   7. A   8. D   9. C   10. D

PART TWO TEST

1. C   2. B   3. D   4. D   5. B

6. D   7. D   8. B   9. A   10. D

11. A   12. B   13. C   14. D   15. B

16. D   17. C   18. A   19. C   20. D

21. B   22. C   23. C   24. B   25. D

26. C   27. D   28. B   29. A   30. C

31. D   32. B   33. C   34. A   35. B

36. A   37. B   38. B   39. C   40. D

CHAPTER 8

1. D   2. C   3. A   4. B   5. D

6. C   7. A   8. C   9. C   10. B

CHAPTER 9

1. B   2. D   3. A   4. D   5. A

6. B   7. D   8. C   9. C   10. B

CHAPTER 10

1. B   2. C   3. D   4. C   5. C

6. A   7. B   8. D   9. A   10. D

CHAPTER 11

1. C   2. D   3. B   4. D   5. C

6. B   7. A   8. A   9. B   10. D

PART THREE TEST

1. B   2. C   3. C   4. A   5. D

6. D   7. C   8. B   9. C   10. B

11. C   12. D   13. A   14. D   15. D

16. D   17. C   18. B   19. D   20. B

21. C   22. B   23. D   24. A   25. C

26. A   27. C   28. C   29. D   30. C

31. B   32. C   33. D   34. D   35. C

36. C   37. A   38. B   39. D   40. A

CHAPTER 12

1. C 2. C 3. D 4. C 5. D

6. B 7. C 8. A 9. B 10. A

CHAPTER 13

1. D 2. B 3. A 4. A 5. B

6. B 7. C 8. D 9. C 10. C

CHAPTER 14

1. D 2. B 3. C 4. B 5. D

6. C 7. A 8. D 9. A 10. B

PART FOUR TEST

1. C 2. B 3. D 4. D 5. A

6. A 7. B 8. B 9. C 10. B

11. A 12. B 13. C 14. D 15. C

16. D 17. C 18. B 19. D 20. D

21. B 22. C 23. D 24. A 25. C

26. B 27. B 28. C 29. C 30. A

31. B 32. C 33. A 34. D 35. B

36. D 37. A 38. C 39. C 40. A

FINAL EXAM

1. D 2. A 3. C 4. B 5. B

6. C 7. D 8. A 9. A 10. D

11. C 12. B 13. A 14. B 15. A

16. C 17. B 18. C 19. A 20. B

21. B 22. D 23. B 24. D 25. C

26. A 27. C 28. B 29. D 30. C

31. D 32. A 33. D 34. B 35. A

36. B 37. B 38. D 39. D 40. C

41. B 42. C 43. B 44. D 45. C

46. D 47. B 48. C 49. A 50. B

51. D 52. C 53. A 54. B 55. A

56. D 57. C 58. D 59. A 60. D

61. C 62. C 63. B 64. B 65. D

66. D 67. B 68. C 69. D 70. A

71. D 72. B 73. C 74. B 75. A

76. B 77. C 78. D 79. A 80. C

81. B 82. B 83. A 84. C 85. C

86. D 87. C 88. D 89. B 90. A

91. B 92. D 93. C 94. A 95. D

96. A 97. C 98. B 99. B 100. A

APPENDIX 1

Acronyms and Descriptions

Acronym	Description
API	American Petroleum Institute
ARAR	Applicable Relevant and Appropriate Requirements (cleanup standards)
ASTM	American Society for Testing and Materials
ATSDR	Agency for Toxic Substances and Disease Registry
BTU	British thermal unit—energy required to raise 1 lb. of water 1°F
CAA	Clean Air Act
CBEN	Center for Biological and Environmental Nanotechnology
CEQ	Council on Environmental Quality
CFC	Chlorofluorocarbon—an ozone-depleting refrigerant
CNST	Center for Nanoscale Science and Technology

Nanotechnology Demystified

(continued)

Acronym	Description
CPSC	Consumer Product Safety Commission
CWA	Clean Water Act
DARPA	Defense Advanced Research Projects Agency
DDT	Dichlorodiphenyltrichloroethane—a toxic pesticide
DNA	Deoxyribonucleic Acid—made of phosphates, sugars, purines, and pyrimidines; helix shape; carries genetic information in cell nuclei
DO	Dissolved oxygen
DOD	Department of Defense
DOE	Department of Energy
DOJ	Department of Justice
DOT	Department of Transportation
DRE	Destruction and Removal Efficiency
DWNT	Double-walled carbon nanotubes
EERE	Energy Efficiency and Renewable Energy
EIS	Environmental impact statement
ELF	Extremely low frequency electromagnetic wave (<300 Hz)—emitted by electrical power lines
EMS	Environmental Management System (also see ISO14000)
EPA	Environmental Protection Agency
EREF	Environmental Research and Education Foundation
ERT	Environmental Resources Trust, Inc.
ESA	Environmental Site Assessment
ESI	Environmental Sustainability Index
FDA	Food and Drug Administration
FIFRA	Federal Insecticide, Fungicide and Rodenticide Act
First Third	August 17, 1988, Federal Register (53 FR 31138)—the first of the hazardous waste land disposal restrictions
GCM	Global climate model
GLP	Good laboratory practices
GMO	Genetically modified organism
GMP	Good manufacturing procedures
HazWoper	29 CFR 1910.120—the OSHA / EPA requirement to have all employees trained if they will be handling, managing or shipping hazardous wastes.
HazMat	Hazardous Material

(continued)

Acronym	Description
HSWA	Hazardous and Solid Waste Amendments, 1984
HWM	Hazardous waste management
Hz	Hertz—frequency with which alternating current changes direction
IEA	International Energy Association
kWh	Kilowatt-hour
MSDS	Material Safety Data Sheet (under OSHA)
MW	Megawatt—1000 kilowatts (1 million watts)
MWNT	Multi-walled carbon nanotubes
NASA	National Aeronautics and Space Administration
NCEM	National Center for Electron Microscopy
NEPA	National Environmental Policy Act
NESHAP	National Emissions Standard for Hazardous Air Pollutants
NIOSH	National Institute of Occupational Safety and Health
NIST	National Institute of Standards and Technology
NGO	Nongovernmental organizations—more than 10,000 organizations worldwide linked by ECONET
NNI	National Nanotechnology Initiative
NPDES	National Pollutants Discharge Elimination System
NPL	National Priorities List of Superfund sites
NRC	Nuclear Regulatory Commission
ORD	Office of Research and Development
OSHA	Occupational Safety and Health Administration
OSW	Office of Solid Waste
OTG	Off-the-grid power generation independent of a major power plant
PCB	Polychlorinated biphenyl—used in dyes, paints, light bulbs, transformers, and capacitors
PSP	Point source pollution
PEL	Permissible exposure limit
pH	Logarithmic scale that measures acidity (pH 0) and alkalinity (pH 14); pH 7 is neutral
ppb	Parts per billion
ppm	Parts per million
PV	Photovoltaic device—generates electricity through semiconducting material

(continued)

Acronym	Description
PVC	Petrochemical formed from toxic gas vinyl chloride and used as a base in plastics
Rad	Radiation Absorbed Dose—amount of radiation energy absorbed in 1 gm of human tissue
R&D	Research and Development
rDNA	Recombinant DNA—new mix of genes spliced together on a DNA strand; (a.k.a. biotechnology)
Rem	R-roentgen equivalent man—biological effect of a given radiation at sea level is 1 rem.
RNA	Ribonucleic Acid—formed on DNA and involved in protein synthesis
SDWA	Safe Drinking Water Act of 1974
SWNT	Single-walled carbon nanotubes
TOC	Total Organic Carbon
TSCA	Toxic Substances Control Act of 1976, which regulates asbestos, PCBs, new chemicals being developed for sale, and other chemicals
TSDF	Treatment, storage, or disposal facility (permitted hazardous waste facility)
TWC	Third-world countries
USDA	United States Department of Agriculture
USGS	United States Geological Survey—manages LandSat which images the environment via satellite
UV	Ultraviolet radiation from the sun (UVA, UVB types)
VOC	Volatile organic compound—carbon-containing compounds that evaporate easily at low temperatures
W	Watt—unit of electrical power

APPENDIX 2

Companies and Products

Company	Products
Acadia Research Corp.	Gene discovery, molecular characterization of disease
Advanced Nano Coatings, Inc.	High performance, VOC (volatile organic compound) compliant epoxy coatings
Advanced Nano Products	Nanocrystalline powders, ceramic targets for sputtering, and e-beam evaporation,
Altair Nanotechnologies Inc.	Lithium titanate spinel electrode nanomaterials
Applied Nanofluorescence, LLC	Optical instrumentation for nanotube study
Applied Nanoworks	Nanomaterials and quantum dot solutions
Arryx, Inc.	Nano-tweezers to pick up and move nanoparticles

(continued)

Company	Products
Aspen Aerogels	Highly-insulative, nanoporous aerogel shoe inserts for cold or military use
BASF	Building materials, hydrophobic coatings
California Molecular Electronics Corp.	Invent, acquire, assimilate, and utilize intellectual property in the field of molecular electronics
Carbon Nanotechnologies, Inc.	Commercial production of carbon nanotubes (Buckytubes)
Cima Nanotech, Inc.	Fine, ultra-fine and nanosized metal and alloy powders
Dendritech, Inc.	Dendrimer manufacturing and production
Dendritic NanoTechnologies Inc.	Branching molecules (dendrimers)with high application diversity (e.g., pharmaceuticals to treat disease, tissues, and organs)
EnviroSystems	EcoTru hospital-grade, nanoemulsive disinfectant cleaner
eSpin Technologies, Inc.	Polymeric nanofiber manufacturing technology
Front Edge	Super thin rechargeable battery
Helix Material Solutions, Inc.	Single- and multi-walled carbon nanotubes
Hysitron	Research/industrial instruments to measure nanoscale strength, elasticity, friction, wear, and adhesion
Integran	Nanocrystalline metals, nano-coatings, nanopowders
Intematix	Phosphors
Intematix Corp.	Electronic materials; catalysts for fuel cell membranes
International Carbon, Inc.	Carbon nanostructures
Kereos Inc.	Therapeutic nanoparticles and imaging tags to target disease
Lumera	Polymer materials and products
Luna Innovations, Inc.	Hollow molecules of carbon atoms that enclose various metal and rare earth elements
Meliorum Technologies, Inc.	Luminescent materials (silicon, metal sulfides, and selenides), metal or metal alloy nanomaterials
Metal Nanopowders Ltd.	Metal nanopowders
MetaMateria Partners LLC	Nanopowders and components for fuel cells, batteries, membranes, other electrochemical devices, filters, rocket nozzles, and catalyst supports
Molecular Electronics Corp.	Electronics and optoelectronic applications
Molecular Imprints	Nanoimprint tool maker for semiconductor and electronics industry
Nano Electronics	New materials like high k gate dielectrics, metal gates and silicides

(continued)

Company	Products
Nano-C Inc.	Manufacture of high-purity fullerenes and other nanoscale fullerenic materials
Nanocor	Nanoclay and nanocomposites for plastics
Nanocs	Water soluble single- and multi-walled carbon nanotubes, biofunctional nanoparticles and coatings
NanoDynamics	Nano-silver, copper, nickel particles; nano-oxides, nanostructured carbons
Nanofilm	Coating technology for sunglasses and windshield; repels water and keeps tar from sticking
Nanogate Technologies	Anti-adhesive and antimicrobial coatings
NanoGram Corp.	Chemical compositions and slurries for computer chips
Nanohorizons	IP portfolio license from Penn State on manufacturing methods of thin-film nanostructures
NanoInk Inc.	Anthrax detection
NanoOpto	Nano-structures for optical system building blocks
Nanophase Technologies, Corp.	Preparation and commercial manufacturing of nanopowder metal oxides
Nanopoint	Allows scientists to peer inside/test living cells at resolutions of ≤ 50 nm. (e.g., IR, visible, and UV ranges)
NanoProducts	Nanoscale powders, dispersions, and powder-based products, (single metal, multi-metal and doped oxides)
Nanosize Ltd.	Specific surface area powders and nano dispersion materials
Nanospectra Biosciences	Non-invasive medical therapies with nanoshells
NanoSpectra Biosciences, Inc.	Non-invasive medical therapies using nanoshell particles
Nanosphere	Analysis and ultra-sensitive detection of nucleic acids and proteins
Nanosys Inc.	Flexible/thin-film electronics, biological substrates, solar cells
Nano-Tex	Nanotech-enabled fabric enhancements/coatings
Nanotherapeutics, Inc.	Nanometer-scale particle delivery for pharmaceutical and over-the-counter products
Nanova, LLC	Nanopowders for paint, coatings, plastics, paper, adhesives, sealants, cosmetics, wire and cable and the health care industries
Nanox	Environmental nanocrystal catalysts
NEI Corp.	Proprietary and patented technologies for making nanopowders and nanostructured intermediates

(continued)

Company	Products
Neo-Photonics Corp.	Nano-optical component maker
Novation Environmental Technologies	License for NASA's iodine-based nanofiltration/disinfection for water purification
Ntera	Electronic ink and digital paper
Nyacol Nano technologies, Inc.	Inorganic metal oxides and organic based silica solutions for flame retardants, abrasion resistance and binders for catalysts, refractories, and ceramic fibers
Oxane Materials, Inc.	High temperature stable PEM fuel cell membranes, nanoparticles, composites, and coatings
Powdermet, Inc.	Metal and ceramic nanoengineered fine powders and particulates
pSivida Ltd.	Developing biosilicon for health care applications
Q Chip	Microfluidic technology for manufacture of micro and nano particles for pharmaceutical, food and cosmetics industries
QuantumSphere, Inc.	Metallic nanopowders for aerospace, defense, energy, automotive and other material application markets
Sensicore Inc.	Lab-on-a-chip multi-sensor devices for water testing and monitoring
Solaris Nanosciences	Rechargeable dye-sensitized solar cell
Starfire Systems	Anti-corrosion and anti-wear nano-structured silicon-carbide ceramics/polymers
Starpharma	Licensing of dendrimer-based pharmaceutical applications
Technanogy	Producer of high-quality, highly energetic ultra-pure aluminum nanoscale powder
Therafuse Inc.	A skin patch for diabetics using nanostraws
Wilson	Nanomaterials in golf clubs and balls, tennis rackets
Zyvex	Nanotube manipulators; nanomaterials

APPENDIX 3

References

J. Aizpurua, et al. 2003. "Optical Properties of Gold Nanorings". *Physical Review Letters* 90, 57401.

Atkinson, W. 2003. *Nanocosm*. New York: Amacom.

Bennett, J., et al. 2003. *London's Leonardo: The Life and Work of Robert Hooke*. New York: Oxford University Press.

Boisseau, P. 2005. "Bringing Nanobio to Life in Europe". *Small Times Magazine* Vol. 5, No. 7, 10–11.

Campbell, C., and J. Laherrere. 1998. "The End of Cheap Oil." *Scientific American* 278 (3), 78–83.

Crandall, B. C., ed. 1996. *Nanotechnology: Molecular Speculations on Global Abundance*. Cambridge, Massachusetts: MIT Press.

Deffeyes, K. 2001. *Hubbert's Peak: The Impending World Oil Shortage*. Princeton, New Jersey: Princeton University Press.

Endo, M., et al. 2005. "Nanotechnology: 'Buckypaper' from coaxial nanotubes." *Nature* 433, 476 doi: 10.1038/433476a.

Energy Information Administration, Office of Integrated Analysis and Forecasting. *2004. International Energy Outlook 2004*. Rep. no. DOE/EIA-0484(2004). Washington, D.C.: U.S. Department of Energy.

Environmental Protection Agency. 2002. *In Brief: The U.S. Greenhouse Gas Inventory*. Washington, DC.: Office of Air and Radiation (EPA 430-F-02-008)

Ferrari, M. 2005. "Cancer Nanotechnology: Opportunities and Challenges". *Nature Reviews: Cancer* Vol. 5, 161–171.

Feynman, R. 1960. "There's Plenty of Room at the Bottom". Engineering and Science Caltech presentation. http://www.zyvex.com/nanotech/feynman.html.

Fritzsche, W., ed. 2002. *DNA-based Molecular Construction*. International Workshop on DNA-based Molecular Construction, Jena, Germany; 23–25 May 2002. Melville, New York: American Institute of Physics, AIP Conference Proceedings, Vol. 640.

Goodsell, D.S. 2004. *BioNanotechnology: Lessons from Nature*. Hoboken, New Jersey: Wiley-Liss, Inc.

Goodstein, D. 2004. *Out of Gas: The End of the Age of Oil*. New York: W.W. Norton & Company.

Hood, L., et al. 2004. "Systems Biology and New Technologies Enable Predictive and Preventative Medicine." *Science* Vol. 306, 640–643.

Jones, R.L. 2004. *Soft Machines: Nanotechnology and Life*. Oxford, U.K.: Oxford University Press.

Lüsted, M. and G. Lüsted. 2005. *A Nuclear Power Plant*. New York: Lucent Books.

Marx, V. 2005. "Molecular Imaging." *Chemical and Engineering News* Vol. 83, No. 30, 25–36.

Melosh, N. et al. 2003. *Journal of Cellular Biochemistry* Vol. 87, pp. 112–115.

Mitlin, D., V. Radmilovic, U. Dahmen, and J.W. Morris, Jr. 2001. "Precipitation and Aging in Al-Si-Ge-Cu." *Metallurgical and Materials Transactions* 32A, 197–199.

O'Neal, D.P., et al. 2004. "Photo-thermal tumor ablation in mice using near infrared-absorbing nanoparticles." *Cancer Letters* Vol. 209, 171–176.

Patolsky, F., et al. 2004. "Electrical Detection of Single Viruses". *Proceedings of the National Academy of Sciences, 28 September 2004,* Vol. 101, No. 39, 14017–14022.

Poole, C., Jr., and F. Owens. 2003. *Introduction to Nanotechnology*. Hoboken, New Jersey: Wiley-Interscience, John Wiley and Sons.

Ratner, M., and D. Ratner. 2002. *Nanotechnology: A Gentle Introduction to the Next Big Idea*. Upper Saddle River, New Jersey: Prentice Hall.

Robichaud, C., D. Tanzil, U. Weilenmann, and M. Wiesner. 2005. "Relative Risk Analysis of Several Manufactured Nanomaterials: An Insurance Industry Context." *Environmental Science and Technology*, 4 October 2005.

Roco, M., and W. Bainbridge. 2001. *Societal Implications of Nanoscience and Nanotechnology*. Boston, Massachusetts: Kluwer Academic Publishers.

Shellenberger, M., and T. Nordhaus. 2004. "The Death of Environmentalism: Global Warming Politics in a Post-Environmental World." Essay presented at the October 2004 meeting of the Environmental Grantmakers Association.

Stix, G. 2001. "Little Big Science." *Scientific American.com* (16 September 2001).

Stockman, M., and D. Bergman. 2003. "Quantum Nanoplasmonics: Surface Plasmon Amplification through Stimulated Emission of Radiation (Spaser)." American Physical Society, Annual APS, 3–7 March 2003, abstract #S11.009.

Watson, J., and F. Crick. 1953. "Molecular Structure of Nucleic Acids: A Structure for Deoxyribose Nucleic Acid." *Nature*, 25 April 1953, Vol. 171, 737–738.

Zandonella, C. 2003. "The Tiny Toolkit." *Nature* Vol. 423, 1 May 2003, 10–12.

Internet References

BUSINESS	• NanoBusiness Alliance: http://www.nanobusiness.org
	• Technology Review: http://www.technologyreview.com
	• International Small Technology Network, Nanotechnology.com: http://www.nanoinvestornews.com/index.php
	• Cientifica Business Information and Consulting: http://www.phantomsnet.com
	• "Technology Topics for Investors" by David J. Roughly: http://www.smallcapmedia.com/pdf/Nanotechfinal.pdf
CHEMISTRY	• WebElements periodic table: http://www.webelements.com
	• EnvironmentalChemistry.com site on periodic table: http://environmentalchemistry.com/yogi/periodic/Pb.html
	• "There's Plenty of Room at the Bottom," by Richard Frynman: http://www.zyvex.com/nanotech/feynman.html
	• University of California, Berkeley, Alvisatos Group: http://www.cchem.berkeley.edu/~pagrp/
COMMUNICATIONS	• National Institute of Standards and Technology: http://www.nist.gov/
	• NEC Laboratories America: http://www.nec-labs.com/
FOSSIL FUELS	• U.S. Geological Survey World Petroleum Assessment 2000: http://greenwood.cr.usgs.gov/energy/WorldEnergy/DDS-60

GEOTHERMAL	• Sandia National Laboratories geotherman Research: http://www.sandia.gov/geothermal • USGS Information on Plate Techtonics: http://geology.er.usgs.gov/eastern/tectonic.html
ELECTRONICS	• Intel Corporation, "Moore's Law": http://www.intel.com/technology/silicon/mooreslaw/ • Online Insider's "IBM, Partners Creating 1,000+ Jobs with $2.7 Billion in New York Projects": http://www.conway.com/ssinsider/bbdeal/bd050117.htm • European Nanoelectronics Initiative Advisory Council: http://www.cordis.lu/ist/eniac/ • NewScientist.com "Nano-transistor self assembles using biology: http://www.newscientist.com/article.ns?id=dn4406 • Snapshots of Science & Medicine, " DNA Chips": http://science.education.nih.gov/newsnapshots/TOC_Chips/Chips_RITN/How_Chips_Work_1/how_chips_work_1.html • IBM, "IBM scientists make breakthrough in nanoscale imaging": http://domino.research.ibm.com/comm/pr.nsf/pages/news.20040714_nanoscale.html
ENVIRONMENT	• Environmental Protection Agency Nanotechnology page: http://es.epa.gov/ncer/nano/ • Enviroene: http://es.epa.gov/neer/publications/nano/nanotechnology4-20-04.pdf • National Center for Environmental Research: http://es.epa.gov/ncer/publications/nano/index.html • Environmental Protection Agency Emergency Response System, "Legal Authorities Defining Hazardous Substances": http://www.epa.gov/superfund/programs/er/hazsubs/lauths. htm
GREEN ENERGY	• Department of Energy, "Energy Efficiency and Renewable Energy": http://www.eere.energy.gov
MATERIALS	• Nanovip.com: http://www.nanovip.com/directory/Materials/index.php • Forbes.com, "The Top Ten Nanotech Products of 2003": http://www.forbes.com/2003/12/29/cz_jw_1229soapbox.html • Wilson: http://www.wilsongolf.com • Larta, "Nano Republic Award Winners": http://www.larta.org/lavox/articlelinks/2003/030721_nanoawardwinners.asp

MEDICINE	• Office of Portfolio Analysis and Strategic Initiatives, Nanomedicine overview: http://nihroadmap.nih.gov/nanomedicine/index.asp • U.S. Food and Drug Administration Nanotechnology: http://www.fda.gov/nanotechnology • Proceedings of the National Academy of Sciences, "Nanocrystal Targeting in vivo": http://www.pnas.org/cgi/content/full/99/20/12617
MICROSCOPES	• Microscopes: http://www.cas.muohio.edu/~mbi-ws/microscopes/index.html
NANOTECHNOLOGY	• Richard E. Smalley Institute for Nanoscale Science and Technology at Rice University: http://www.cnst.rice.edu • National Nanotechnology Initiative: http://www.nano.gov/ • Nanotechweb.org: "Drexler dubs 'gray goo' fears obsolete": http://nanotechweb.org/articles/society/3/6/1/1 • Nano Science and Technology Institute: http://www.nsti.org • Lux Research: http://www.luxresearchinc.com/ • Nanotechweb.org, nanotechnology news and events: http://www.nanotechweb.org/
NANO NEWS	• SmallTimes news: http://www.smalltimes.com • Foresight Nanotech Institute: http://www.foresight.org/nanodot/ • Institute of Nanotechnology: http://www.nano.org.uk • National Institute of Standards and Technology: http://www.nist.gov/
OCEANS	• U.S. Geological Survey: http://www.usgs.gov • National Oceanic and Atmospheric Administration: http://www.nws.noaa.gov
SOLAR	• U.S. Department of Energy Solar Energy Topics: http://www.eere.energy.gov/RE/solar.html
SPACE	• NASA: http://www.nasa.gov/home/index.html

INDEX